A Brief History of Cells

A Brief History of Cells

Cheng Lin
Translated by Feng Ye

naturalogic

A Brief History of Cells

Written by Cheng Lin
Translated by Feng Ye

Copyright © Shanghai Jiao Tong University Press, Cheng Lin, 2025

上海交通大學 出版社
SHANGHAI JIAO TONG UNIVERSITY PRESS

Published by Naturalogic Publishing Inc.,
under an exclusive license with Shanghai Jiao Tong University Press.

First English Edition 2025
ISBN: 978-1-4878-1301-7

naturalogic | NATURALOGIC
Passion is the genesis of genius.
www.naturalogicpublishers.com

19-1235 Johnson St. Coquitlam, BC, Canada V3B 7E2

Contents

Contents

Preface

When my colleagues and friends ask me why I wrote this book, I usually tell them that it grew out of questions asked by two generations—the old and the young. The old refers to my father, a veteran who was a doctor in the army. He went there to help with the Tangshan earthquake and ran a clinic for a few years after he was discharged, so he knows a bit about medicine. Since I have been doing research in the hospital, he has often asked me what I am working on. When I tell him that I am working on cell reprogramming or, in other words, the research on how to change the fate of cells, he always looks puzzled. Then I give him common examples, such as turning skin or urine into blood. Based on his experience and knowledge, he takes it as a fantasy and is full of worries about my future research. The young ones are my nephew and daughter, who are at the most thoughtful age. When they ask me what I do and I repeat the answer I gave my old father, they are always full of curiosity and astonishment with more questions about how it could be achieved coming up. It's not something that can be answered in one or two sentences. When I tried to search for some related popular science books on the market to answer their questions,

I found these books were either too specialized or too childish. So I decided to write a book myself. But that was easier said than done. I had no idea how to start or what to write about, and the work at the institute was so intense that it was delayed for several years. On the fifth anniversary of the founding of my research group, I invited my students to a dinner to celebrate the occasion. When we talked about the chaos of cell therapy in the current market, the topic moved to our social responsibility as researchers. I didn't think too much about it and blurted out the phrase, "write a reliable book for science popularization," in front of my students. Afterward, I immediately felt that I had blown my own horn, but as a teacher, I couldn't go back on my word. So, I started to conceive and write the book. In retrospect, this invisible pressure pushed me to go ahead on my own, which is not a bad writing strategy for a lazy guy.

Coming from a biology background, I am familiar with cell biology. Therefore, this book is written with great reference to the content of traditional cell biology courses at university but with much of the technical content omitted. Instead, I use simple analogies to explain each technical concept and make it easier for readers to understand. At the same time, and perhaps more remarkably, the book delves into the history of the development of these concepts. In addition, current hot topics about cells, such as stem cells and cell therapy, especially tumor immunotherapy and closely related gene therapy, are introduced. Plant cells, which are on par with animal cells but have long been neglected, are also included. The discovery and study of cells cannot be separated from the microscope from beginning to end, and the history of the development of the microscope is also very important; as the saying goes, "To do a good job, an artisan needs the best tools." These stories are also briefly introduced at the end.

Cells are the smallest unit of a living organism, and the discoverer of every cell, organelle, or cellular phenomenon is also an ordinary person. Both are a drop in the ocean of the world and the long river of history, just like the dust of life. However, behind every grain of dust, there is a story. These stories, big and small, form the main body of this book. It shows not only the hardships and history of research on cells but also the magic and value of discovery.

I hope this book can enable readers, whether old, young, teenage, or middle-aged, to understand the big events in cell science. It will help you recognize the ordinary life of scientists, understand that cell science is not lofty, that the discovery of cells comes from the bits and pieces around us, and that the application of cells affects our lives, thus arousing readers' curiosity about cells.

Cheng Lin

Professor, PhD supervisor, and Principal Investigator
in Ruijin Hospital affiliated to Shanghai Jiaotong University School of Medicine,
Shanghai Institute of Hematology

Discovery of Cells

$\mathcal{M}$ore than 300 years ago, in the seventeenth century, the world was in the midst of the late Renaissance. For hundreds of years before that, almost the whole of Europe had been under the oppressive rule of the Church. People had long since grown tired of that life and had begun to long for the freedom of thought introduced by ancient Greece and Rome. They attempted to revive the culture and art of that time and initiated the movement known as the Renaissance. However, the inquiring minds were not satisfied with the simple inheritance of the ideas of the philosophers. Although the philosophical ideas of Plato, Aristotle, and others had led people to recognize and think about the world to a great extent, they were still discursive in nature and lacked solid evidence. Under the influence of new trends of thought and the development of technology, people gradually began to move from abstract thinking to empirical observation. They tried to establish an objective understanding and description of nature, thus opening the door to modern scientific experimentation.

The Renaissance began in Italy and quickly spread across Europe with the culture of free thought. The activation of a large number of people who loved art and explored science also produced many great scientists, literary figures, and artists who have influenced history. Among these scientific explorers, there were both scientists who loved literature and art, as well as artists who loved science. In the early days of modern science, unlike today, there was no clear distinction between different specialties. Rather, there was a blending of them, resulting in a number of generalists who had never been seen before and might never be seen again. Leading figures of this period and their achievements are still well-known and memorized in history.

In the field of astronomy, Polish priest and astronomer Nicolaus Copernicus established the heliocentric theory, which replaced the original geocentric theory. This theory proposed that the center of the universe was the Sun, not the Earth. Although both theories are wrong from today's perspective, based on what was known at the time, the proposal of heliocentrism is still a huge scientific advance. The German astronomer Johannes Kepler built on the heliocentric theory by stating that the planets moved in elliptical orbits. After them, Galileo Galilei, an Italian astronomer and physicist, was the first to use an astronomical telescope to observe the night sky and discovered the moons of Jupiter. Considering that the Earth also had the Moon as a moon, he believed that the Earth was also just a planet that could move. However, his view was opposed and rejected by the Pope at the time. For this reason, he had to be placed under house arrest for the rest of his life in order to keep his discovery.

In the field of physics, the most famous figure is Isaac Newton, the English physicist and mathematician who solved the riddle of the free fall of an apple and published his biographical work, *The Mathematical Principles of Natural Philosophy*, in 1687. Differences in language created communication barriers. To facilitate communication, Latin was adopted as the common language of the time for professionals, similar to English today. This was also the case with Newton's masterpiece. However, as a collection of relatively cryptic mathematical principles described in a specialized language, few people at the time were able to understand the book, and it was not until 1729 that an English translation of the book appeared. It is said that when Albert Einstein later proposed the theory of relativity, few people could understand it either. No wonder it is often said that the truth is always in the hands of a few. One of the reasons may be that the truth is too difficult for most people to read and comprehend, so naturally they do not understand.

In humanities, Italy is represented by Dante Alighieri, the poet of *The Divine Comedy*; Leonardo da Vinci, the painter of the *Mona Lisa*; and Michelangelo Buonarroti, the sculptor of *David*. Spain is represented by Miguel de Cervantes Saavedra, the author of *Don Quixote*. England is represented by the household name of William Shakespeare, whose masterpieces are too numerous to list, including the popular love drama *Romeo and Juliet* and *Hamlet*, one of the four great tragedies.

Europe is in its heyday after the Renaissance, and with all kinds of new ideas and discoveries competing for openness and splendor, the Netherlands is not to be outdone. It ushers in an unprecedented development. The Netherlands is located in the western part of Europe, between the United Kingdom and Germany, with Belgium to the south and Amsterdam as its capital. About 50 kilometers south of the capital is the city of Delft, close to The Hague and the hometown of Johannes Vermeer. Like other cities in the Netherlands, Delft still retains its old canals and arched bridges. On both sides of the river are tightly packed houses with flowers and plants planted in front of them. They are fresh and elegant, reminiscent of the ancient city of Lijiang in Yunnan Province, China. Even today, when you set foot in Delft, you can feel the centuries of history in the buildings, streets, restaurants, bars, and cafés. All of these have a distinctly European flavor. Thanks to the development of navigation promoted by the Renaissance, Chinese blue and white porcelain became popular in Europe, leading to the development of ceramics, pigments, and painting industries in Delft. The merchant Johannes Vermeer struggled to support his eleven children by running a small hotel and selling paintings. In addition to this, he also tries to paint himself. But unlike other painters at that time, he drew on the then-developing theory of optical perspective, particularly making extensive use of dark boxes. This gave his paintings a distinctive three-dimensional perspective of light and shadow. Unfortunately, his paintings were not recognized by the world during his lifetime, and it was not until more than 200 years after his death that they began to be appreciated. As a result, not many of his works have survived. This is partly because they were not adequately protected and partly because he was not a prolific painter. Of the more than 30 works that have survived, the best-known are *Landscape with Delft*, *The Maid Pouring Milk*, and *The Maiden with a Pearl Earring*.

Delft, Netherlands

Another person who lived in the same time and city as Vermeer was the subject of our next feature. He was born on October 24, 1632, in a brick house with a pitched roof on the corner of the street known as the Lion's Gate. His grandfathers, parents, uncles, and aunts lived there at the time. Initially, the protagonist's nickname was Thonis Philips, and his surname was derived from the transformation of the name of his birthplace, Levenhuk, which means "from the lion's corner." Years later, the protagonist changed his name to Antonie and added van to his middle name, thus becoming known to posterity as Antonie Philips van Leeuwenhoek. It has been suggested that Vermeer and Leeuwenhoek were good friends, especially since Leeuwenhoek helped to manage Vermeer's estate after his death. Additionally, it is believed that Vermeer's technique of drawing in the dark may have originated with Leeuwenhoek's help. However, there is no evidence of direct communication between the two men, either in the form of letters or a cup of afternoon tea in the same café.

Therefore, it is more likely that they were just passers-by who never said hello to each other.

Antonie Philips van Leeuwenhoek

Leeuwenhoek's family consisted of seven siblings, and he lost two of them and his own father when he was six years old. Two years later, his mother remarried with five children. His stepfather was a 60-year-old painter. Leeuwenhoek's family was Christian, and although many Christians at the time sent their children to public schools, Leeuwenhoek's mother sent him to a Protestant school at the age of eight. There, he learned simple Latin. At the age of fourteen, Leeuwenhoek began to live with his uncle, who was a sort of magistrate. From him, he learned the basic rules of the law. Two years later, at the age of 16, his mother sent him to Amsterdam to learn the apparel trade. Life in the capital was colorful, but as an apprentice, Leeuwenhoek had no time to enjoy it. In order to earn his own living in the future, he had to follow his master's instructions and learn how to order, negotiate, and file tax returns. It was during this time that Leeuwenhoek was first introduced to the magnifying glass, a type of lens with two convex sides used to observe the quality of textiles.

The discovery and use of convex lenses dates back to before the Common Era. They were used primarily to aid in carving and to focus light for ignition. After the eleventh century, they were gradually used to magnify text for easier reading. It was not until the second half of the thirteenth century that lenses were mounted in metal frames, forming the basis of spectacles to improve the vision of the elderly. This was followed by the use of concave lenses to improve the vision of short-sighted people. By this time, large and small opticians had appeared in various European countries. In addition, some people began to try to fit several lenses into the same tube to observe things at a greater distance. However, there is no way of proving who was the first person in history to make a telescope. It was also during this period that Galileo began to use the telescope to observe the stars, modifying it to greatly improve its performance and thus making his later great discoveries. Although there is evidence that Galileo also made microscopes shortly afterward, it is not known whether he was the first to do so. The impetus for the widespread popularity of the microscope is attributed to Cornelis Drebbel, an instrument maker of the time who was also an inventor and engineer. He designed and built the first submarine in history.

Six years had passed since Leeuwenhoek came to Amsterdam to learn from his master. Finally, after a long absence, he returned to his hometown and, at the age of 22, married his wife. Together, they ran their own clothing shop. They soon bought a house of their own and raised three sons and two daughters. However, only one daughter, Maria, grew up to accompany Leeuwenhoek to the end of his life, while all the other children died young. As the microscopes sold on the market either did not have enough magnification or were defective in various ways and could not satisfy the need to check the quality of the cloth, Leeuwenhoek decided to make his own microscope. It is estimated that he made more than 500 microscopes during his lifetime. Only eight and a half have survived. Why is there half a microscope left? While the other eight microscopes are all intact, including lenses and stands, the lens on the ninth one is missing, leaving only an empty stand. According to modern technical measurements, the smallest magnification of the eight undamaged microscopes was 69 times, with a maximum magnification of 266 times. Early microscopes were relatively simple, consisting of a metal tray or stand with a glass bead or tiny lens embedded in it. But for Leeuwenhoek to be the best, he not only burned the glass beads with fire and polished the lenses with sand in person but even handmade the metal stand with smelting and casting. From this, we can see his attitude of excellence. It is not difficult to understand why he could use the microscope to discover the world that others ignored. Although the microscope had been invented decades earlier, most people only used it for recreational observation.

Nearly 30 years old, Leeuwenhoek secured a position in the government, and his letter of appointment is still preserved in the city diary of Delft. As a civil servant, he earned a steady income that enabled his family to prosper. This economic base, together with the rise of science and his fascination with and mastery of the microscope, stimulated Leeuwenhoek's interests. No longer content with merely observing the threads of silk in cloth, he used a microscope to observe everything he could gather around him, even insects. Great discoveries often depend on the development of technology, and the observation of tiny organisms is no different. Using these advanced microscopes, which were unique at the time, Leeuwenhoek collected a large number of microscopic anatomical drawings of insects such as flies, moths, and worms. By the time he had microscopic atlases of a number of specimens, Leeuwenhoek had made the transition from amateur to professional scientist. No longer content with self-indulgence, he was eager to be recognized by other experts in this field.

At that time, the Renaissance was giving rise to a group of natural philosophers. Francis Bacon, one of the leading figures, believed that scientific knowledge should be based on objective observation and rigorous experimental verification. He also believed that generalization, summary, and analysis of objective knowledge were indispensable. He is, therefore, regarded as the pioneer of English materialism, which was greatly appreciated by Karl Heinrich Marx. However, although Bacon's ideas were in line with the trend of history, he was only a philosopher and did not really carry out any experimental scientific research. He was, at most, a preacher of science. It was the Royal Society of London that really took up his ideas and put them into practice. On December 28, 1660, the Royal Society of London was founded on the initiative of 12 gentlemen, including Robert Boyle, John Wilkins, Robert Moray, and Viscount Brouncke, with the aim of exploring nature and analyzing the scientific laws behind everything through scientific experiments. It was these experiments that started modern scientific research in physics and mathematics.

In order to promote the healthy growth, the society has formulated three principles: First, a regular meeting should be held every Wednesday to discuss the content of the experiments to be carried out and the direction of future scientific research; second, they should have their own journal to publish the scientific progress of its members in order to promote exchange, which was not only circulated in Britain but also began to spread to other European countries, including the Netherlands; third, they need to recruit an administrator whose duties would include applying for government funding and carrying out substantial and concrete experiments on scientific topics discussed by its members. Under this heading, our second protagonist finally came into his own, and he was Robert Hooke.

Royal Society of London

Robert Hooke

To understand why Hooke was able to serve as the administrator of the first scientific society in history, one must begin with his personal history. Born on July 18, 1635, in a small village on the Isle of Wight, England, Hooke was sickly from birth. His parents never sent him to school as a young man, and he was entirely self-taught, or in other words, self-indulgent at home. There, he busied himself dismantling and reassembling various mechanical devices. His father was also in poor health, suffering from various illnesses, and died when he was 13. Hooke had to leave his hometown and move to London to earn a living, starting as a laboratory boy at Westminster School. Here, he learned Latin and Greek in a systematic way and was fortunate enough to meet the famous S. Wilkinson and Boyle. The latter, in particular, held Hooke in high esteem, and it was with Hooke's help that Boyle was able to construct the pneumatic pump with remarkable success. It was, therefore, only natural that Hooke, the young man favored by both founders, should be appointed as the first administrator of the society when it was formed.

However, it is not an easy role to take seriously. As an administrator, he had to not only coordinate the members' time to hold weekly meetings but also take on the role of a technician, designing elaborate experiments to test the various strange scientific ideas put forward by the members and trying to satisfy their curiosity as much as possible. Because of the open-ended, almost all-encompassing nature of research in the early days, Hooke had to move between a number of fields, including pendulums, respiration, combustion, magnetic fields, gravity, telegraphy, astronomy, and even music. It was the intersection of these disciplines that contributed to Hooke's ability to thrive in the field of science while sowing the seeds of friction between him and other members. In the field of scientific research, sometimes ideas are important, sometimes experiments are important, and sometimes both are equally important. We know that one of the members, Newton, was the first to suggest that sunlight is made up of seven different colors. However, when he proposed this idea in 1675, he was heavily criticized by Hooke, who had done a similar experiment ten years earlier. For this reason, Newton later had to admit that his discovery had indeed been influenced by Hooke. The two men went from being partners to enemies. In 1687, when Newton published his greatest laws of motion, he made no mention of Hooke, although some people believe that Hooke indeed had a contribution. We all know that Newton famously said, "If I have seen farther than others, it is because I have stood on the shoulders of giants." Whenever this statement is mentioned, we think it is Newton's humble remark about his great discovery. But this is not the case. He said this just to show that his discovery had nothing to do with Hooke as Hooke was a very short man. It is actually an insult and a denial of Hooke.

Of course, Newton's denial did not stop Hooke from exploring other fields. As the old saying goes, gold always shines! Leeuwenhoek was not the only researcher looking at microscopic things under the microscope, and Hooke was one of them. He pioneered the observation of plant cork tissue by cutting and placing it under the microscope. Small chambers were discovered on the tissue with neat and orderly arrangements, which were closely linked to each other so that they were named cells. This was the first time in history that the discovery of cells and their name were truly documented. Interestingly, however, Hooke did not communicate these discoveries with Leeuwenhoek privately, but instead published a collection of papers on his discoveries and microscopic drawings of insects and textiles. *The Micrographia*, published in 1665, was a sensation. It gave the world its first comprehensive view of an amazing microscopic world that could not be observed with the naked eye. This work made Hooke who he was. Although the structure observed by Hooke was not really a cell, but a cavity left by a dead cell, and Leeuwenhoek was the first person to observe a living cell, the discovery of the cell was still attributed to Hooke. Leeuwenhoek could only be regarded as "getting up early in the morning but catching up the fair late."

Leeuwenhoek was once asked if Hooke's book had inspired him to make his own observations. He vehemently denied it and always maintained that he had never read the book. But his claim is hardly convincing. As soon as the book was published, it became popular throughout England. At the time, Leeuwenhoek was visiting his late wife's family in England. It was during this time that he made contact with the Royal Society of London and became acquainted with its journal, the *Philosophical Transactions*. He then corresponded frequently with Hooke and the society, informing them of his microscopic discoveries in the hope of gaining their approval. However, as it was difficult for other researchers to obtain Leeuwenhoek's high-magnification microscope, they failed to verify each of his personal discoveries. Therefore, the society did not place much faith in the results. They also made several unsuccessful requests for Leeuwenhoek's help in building advanced microscopes. As a result, most of Leeuwenhoek's discoveries are confined to his surviving correspondence, and only a few were published in the society's journal.

We often say that scientific discoveries are often born of interest. It was Leeuwenhoek's love of science that led to his obsession with microscopic observation. Although unsuccessful at the time, he was inspired by Hooke's cell theory and turned to trying to observe even smaller organisms. One sunny afternoon, he and his friends went on a trip to a green field beside a lake with their microscopes. They accidentally took a drop of water from the lake to observe. Surprisingly, he found that there were many tiny organisms in the clear water that could not be seen directly with the naked eye. These tiny organisms, which had never been discovered by the world, were swimming in the water. The significance of the discovery of these tiny organisms in a drop of water is absolutely no less than the discovery of cells, which proved to the world for the first time that there was another biological world. Just as the observation of galaxies beyond Earth at that time, it showed that there was a micro-world and a macro-world.

Fascinated by these marvelous microscopic worlds, Leeuwenhoek collected samples from almost every conceivable water source and observed the tiny organisms in them. These included rainwater, gutter water behind the eaves of a house, freshly collected water, and water that had been stored for years. It is perhaps a true reflection of early modern science that large, laborious sample observations were made using simple tools, leading to numerous discoveries and scientific conclusions that have stood the test of time. Through long-term microscopic observation of various water sources, Leeuwenhoek found that sealed distilled water, no matter how long it was stored, did not produce micro-organisms. However, once exposed to the air, microorganisms did appear. This discovery should be regarded as a pioneering experiment in microbiology. He also attempted to treat water containing microscopic organisms with a variety of chemical reagents. After observing, counting, and analyzing the changes in the individual life within it, he found that certain chemicals had the efficacy to significantly inhibit the growth of such microscopic organisms, especially for liquid samples of their own oral origin. It is often necessary to look at things in two ways: the good and the not-so-good. Leeuwenhoek found the microscopic organisms in the water to be wonderful and incredible, but he also found the presence of microscopic organisms in his mouth to be disgusting and was hell-bent on getting rid of them. So, based on his own in vitro experiments, he began to use chemicals that inhibited microscopic organisms in the water in his own mouth rinse. This was, in a sense, the prototype of modern toothpaste. None of these discoveries attracted enough attention at the time, but now they seem so great. Like the artistic treasures that have been passed down through history, they have endured, and the fascination of science has been hidden within them.

After becoming famous, Leeuwenhoek also began to take on apprentices to learn how to make microscopes and observe the microscopic world. Out of curiosity, Leeuwenhoek and his students began to observe the semen of various animals and human, which was found to be tadpole-shaped and able to swim.

This would have been the earliest observation and discovery directed at sperm. However, in later historical accounts and reprints, it is widely believed that Leeuwenhoek simply equated the sperm he observed with tiny humans, citing his descriptions in his letters. In fact, these were merely the conclusions of his students, which were objectively described and rejected by Leeuwenhoek in his correspondence with the Royal Society. This shows that Leeuwenhoek was an extremely rigorous scientist who respected factual observations and maintained a lifelong love of science. Moreover, it was his persistence that made him a late bloomer. Many of Leeuwenhoek's discoveries were unintentional. In addition to microorganisms in water, he was the first to use a microscope to observe the presence of red blood cells in blood. However, while observing a drop of blood squeezed from his thumb, he intended to see if there were any salt particles in it. He did not expect to see numerous red and floating spheres. To verify these interesting findings, he also observed the blood of rabbits. Through extensive observation and statistics, he calculated that the diameter of each red blood cell is 1/30 of an inch (1 inch approximately equals to 2.54 centimeters) or about 8.5 microns, a value very close to the diameter of red blood cells measured by modern science and technology (7.7 microns). From these discoveries, we have learned that there are many instances of scientific discovery where a watched flower never blooms, but an unattended willow grows. But it would be a mistake to attribute science to luck because there is another saying, "Opportunity only comes to those who are prepared." Long-term persistence in a given field may seem tedious, but it is only through this approach that we can capture and freeze those moments that can change time. When the idea and the unexpected arrive, we must be prepared to seize them, or else they will slip away unintentionally.

Having learned about the journey of discovery of the cell, you must be wondering why cells are called or translated as "细胞" (*xibao*) in Chinese. What was the original word for cell? By looking at the original documents, we know that the original word "cell" means a small chamber. To find out why the cell was eventually translated into what is now widely known as *xibao* in Chinese, we need to look at the modern history of China that begins with the Opium

Wars of the late Qing Dynasty (AD 1616–1911) and ends with the founding of the People's Republic of China. Although China was in dire straits during this period and the people did not have enough to live on, this did not stop our aspirants from searching for the truth. The more difficult the time was, the more the potential of the people, especially their indomitable spirit, was stimulated to make contributions no less than in times of peace. With China's gate opened by imperialist ships and cannons, a large number of foreign missionaries poured into China, accompanied by foreign books documenting all kinds of scientific discoveries, including foreign botanical books with descriptions of cells. At that time, there were many versions of the translation of the word "cell." For example, Mr. Sun Yat-sen, in his letter to his son Sun Ke, recommended the book *Cell Intelligence* and translated "cell" as "*shengyuan*" (生元). The person who finally translated the word "cell" into the word "细胞" in Chinese was Li Shanlan.

Li Shanlan

Who is Li Shanlan? According to today's education system, he is definitely a child prodigy. He was born on January 2, 1811, in Haining, Zhejiang Province, a small town between Shanghai and Hangzhou. In modern times, it is known for its leather production, but it has a long history with famous people such as Xu Zhimo, Wang Guowei, and Jin Yong. Li Shanlan had been fond of studying arithmetic since childhood. At the age of 9, he could read *Nine Chapters of Arithmetic*; at the age of 14, he self-studied *Geometry*; and at the age of 30, he began to write books such as *Zeguxi Zhai Arithmetic* and *Method to Discover the Roots of Numbers*, which made him become a famous mathematician and the originator of modern mathematical education in China. At the age of 40, Li came to Shanghai and began translating Western scientific works. He insisted on this translation for eight years and became the first person to translate science and technology in modern China. In 1858, at the London Missionary Society Press in Shanghai, together with British missionaries Alexander Williamson and Joseph Edkins, he translated *Fundamentals of Botany* by British botanist John Lindley into Chinese. This translation was titled *Botany* and became the first modern Western botany text to be translated into Chinese. Chinese translation of modern Western botany, a book of eight volumes totaling more than 35,000 words, in which "细胞" is mentioned for the first time at the beginning of volume II. Interestingly, it has been suggested that Li Shanlan originally translated it as "小胞" (*xiaobao*), but his local dialect often pronounced "*xiao*" (小) as "*xi*" (细), leading to the term "细胞." The word "细胞" hence replaced "小胞" and is still used today. Soon after, the book was introduced to Japan. In Japanese, the word "cell" was also written as "细胞," which further gave rise to the word "细菌" (*xijun*, bacteria) and so on. In modern scientific terminology, many of the Chinese translations in the early days came from Japanese translations, such as atom, molecule, science, and society. However, not many words were translated in China and then imported into Japan. The word "细胞" is one of the few. In addition to these, Li Shanlan also translated and published works such as *The Study of Gravity* in mechanics and *Talking about Heaven* in astronomy. The former was the first translated work on the mechanics of motion and the mechanics of rigid bodies and fluids in the history of modern science in China. The latter introduced concepts such as heliocentrism and gravity to China for the first time. As a mathematician and translator, Li Shanlan made indelible contributions to the development of many fields of modern science in China.

Looking back at the history of the discovery of the cell and its translation, in the rolling tide of history, both the European Renaissance and Chinese inventions helped to increase momentum. Among these things, a change in thinking was the source of all exploration; the use of the compass made voyages no longer a dream; the invention of papermaking and printing meant that knowledge was no longer confined to a small geographical area, but allowed the countries of Europe, Europe, and Asia, and even the world, to connect with each other and share in the marvelous discoveries and advances. The learned generalists, made possible by the intersection and integration of different disciplines, were also able to exercise their talents to the full. In today's China, where disciplines are categorized as specialized, refined, special, and new, it is worth reflecting on the fact that while the basic conditions are getting better and better, breakthrough discoveries are becoming fewer and fewer. We hope to inspire people by looking back at the road traveled in history and using the journey of a small cell as a starting point to get a glimpse of the big picture.

Appearance of Cells

*J*ourney to the West is one of the Four Great Classical Novels in Chinese literature. The TV series adapted from it are the childhood memories of the post-70s and post-80s generations and have become unsurpassed classics. Two of the most memorable characters are Sun Wukong and Zhu Bajie. The former is described in the original text: "Round eyes, peaky ears, a hairy face, a God of Thunder's mouth, skinny looking with sunken cheeks, on a body less than four feet tall, like a monkey eating pine cones; though he has the appearance of a human, he is much thinner and bonier." The description of the latter goes, "When he first came, he was a black fat man. However, later on, he became a dork with a long mouth, big ears, bristles at the back of his head, a rough and fearful body, and a head and face like that of a pig. Yet his stomach was insatiable, eating three to five buckets of rice for one meal." From this, we can imagine their height, weight, and facial appearance. Apart from the characters in myths, there are more than eight billion people on earth. Except for twins and multiple births, there are huge differences between people in terms of appearance, such as height, body shape, facial features, and so on. The saying goes that "no two things in the world are exactly alike," and this holds true for people as well. Of course, when we categorize in general terms, we can find both differences and commonalities among individuals. For example, people can be grouped by various characteristics such as cultural background, language, or profession. Similarly, in the case of cells, while it may be difficult to distinguish subtle differences between individual cells, careful classification reveals that certain types of cells have distinct features, and the differences between types are easy to identify. The idea is that things naturally group according to similarities— people form communities, and cells are classified by type—essentially the same concept.

So, how big is a cell? This is the first question many people ask when they hear the word "cell." When you say that you can't see at all with the naked eye, it's an unimaginably small thing. So how small is it eventually? Here's an intuitive analogy: if a cell is magnified to the size of a sesame seed, then that sesame seed should be magnified to the size of a watermelon, a magnification of one to two hundred times. Compared with the most common hair strands, which have a diameter of about 50 or 60 microns, it is usually about a fifth of the diameter of hair strands, so approximately 10 microns up and down. Our human eyes can observe the minimum distance between the two points is about 100 microns; no matter how close your eyes are to the object or how strong the light is, it cannot significantly improve this resolution limit, which is often referred to as resolution rate. Therefore, our eyes cannot directly observe a cell, and we have to use an optical microscope to get a glimpse of it. However, this is only the usual situation. Due to the different types of cell morphology, there is a significant difference in size. The diameter of the smallest cell is only 5 or 6 microns, while the largest cell diameter is up to more than 200 microns. This is a difference of 40 to 50 times. This brings to mind the largest single-aperture astronomical telescope in China, and indeed in the entire world to date—the FAST radio telescope in Guizhou, which has a diameter of 500 meters, while the ordinary observatory reflector telescope caliber is only a few meters or a dozen meters. The gap between cells and cells is reflected in the gap between telescopes and telescopes, where one can span several large mountains, and one only lives in a hut at a snail's pace.

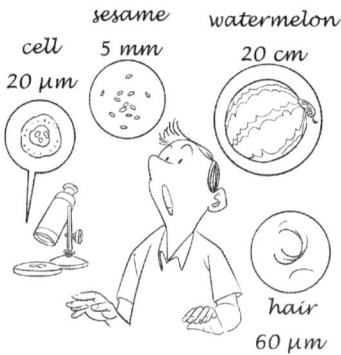

Having said that about size, let's talk about numbers. For an adult weighing 70 kg, there are 20 to 30 trillion cells in the human body. These cells can be divided into more than 200 different cell types depending on the type of tissue. Furthermore, these cells are constantly being replaced. What is this concept? To put it bluntly, it's about finding out how many cells are in a piece of meat. A few years ago, someone started to develop a kind of artificial meat that was synthesized entirely from artificially cultured cells. At that time, they were using cells from cows, and in the end, they got a small piece of steak with more than ten million cells in it.

Following paragraphs present some of the most unusual and distinctive cell types, along with their sizes and quantities, organized from the top to the bottom of the human body. Let's see if your imagination can match the wonders of nature's diversity.

In the human brain, there are about 170 billion cells. Among them, there are more than 80 billion neuron cells, accounting for about half. The remaining part is mainly glial cells, including astrocytes and so on. The shape of neurons is extremely like a small dot radiating lines at different lengths. Short ones among them are called dendrites. Meanwhile, the longest lines can be more than ten times the length of others, and it has a name called axon. These long axons are protected by a myelin sheath that, from a distance, looks like a long chopstick threading together many small sausages. In addition, on each of the thin threads, there grew a number of small projections that very much resembled thorny rose stalks, just like the split ends of hair. However, unlike hair, which normally only divides once into two ends, the axon can have ten or even more divisions. What are these divisions for? Mainly to communicate with other nerve cells. When we watch films about aliens, we often see a scenario where the first contact between a human and an alien is made by extending an arm to each other and opening the fingers. A light touch is made between two creatures strange to each other through the protruding fingertips, which is done to understand the other's purpose and to sense the other's good or bad intentions. Neurons communicate in a very similar way, and that is why these touching ends are called synapses. Imagine the sensation of a synapse when you gently touch the fingertips of your right hand with the fingertips of your own left hand. The sensation is obviously different from a direct handshake, but it is real.

Based on slight differences in their morphology and the functions they perform, neurons can be categorized into many types. But if you divide them by the thickness of the axon, they fall roughly into four categories, from thick to thin, with diameters ranging from more than 20 microns to a few microns. Don't underestimate the difference of a few micrometers. The different thickness determines the speed of signaling in neurons. Sometimes we hear people say that a person is slow to react and that he or she is a nervous wreck. In fact, the opposite is true: the thicker the axon of a neuron, the faster the reaction speed. In humans, the speed of reaction is often reflected in the speed of thinking and movement. The speed of movement, which is reflected in the neurons, mainly refers to the speed of current conduction within and between neurons. For the four types of neurons mentioned above, the speed of current conduction in them is equivalent to that of a fast train, a fast car, a slow bicycle, and a slow walk. Therefore, to say that a person has the feeling of being electrocuted is definitely not a simple description, but a real event that takes place in a neuron.

How was this event hidden in specialized cells discovered? Mainly thanks to two British scientists—Alan Lloyd Hodgkin and Andrew Huxley. In 1939, working together, they inserted two glass needles into the nerve fibers of the squid and discovered for the first time that neurons generate electrical signals when stimulated. They proposed a model named after them, the Hodgkin -Huxley model, which opened up the field of neuroelectrophysiology and has become the mainstream of modern neurobiological research. It was for this work that they were awarded the Nobel Prize in Physiology or Medicine in 1963. The scientist who shared the prize with them was John Carew Eccles, the discoverer of the neuronal synapse. What needs to be explained here is: why did they have to experiment with the squid? Because of the technical conditions at the time, the syringe made by heating and stretching the glass tube was still relatively thick and could not be inserted into an ordinary tiny neuron. But the neurons of squid are different. Under the membrane of its body, squid has incomparably thick neurons. The diameter of its axon is even thicker than the thickest mammalian neurons, ranging from hundreds to thousands of microns. Additionally, the length of it is as long as tens of centimeters. These unique conditions were a dream come true for early neuroscientists, allowing them to carry out experiments in a simple and crude way. Of course, with the development of technology, scientists have now developed membrane clamp technology. No longer do they rely on squid to carry out experiments. Regardless of the size of neurons in any kind of animal, they can use this technology to detect current.

Alan Lloyd Hodgkin   *Andrew Huxley*

There is also another group of cells in the brain that is often overlooked: the glial cells. They look very different from neurons, but they also have lots of bumps, usually a dozen or more. The tip of each bump also produces tiny branches that, from a distance, look like shining stars in the night sky. This is why glial cells of this shape are called astrocytes.

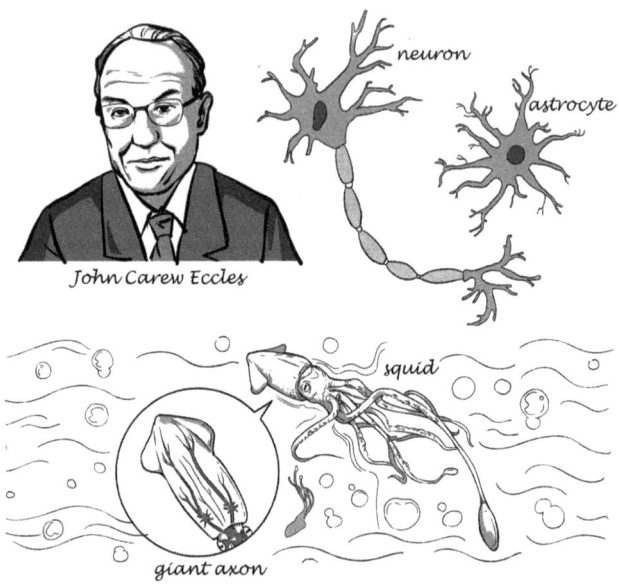

John Carew Eccles

neuron

astrocyte

squid

giant axon

It is often said that the eyes are the windows to the mind, and the mind does indeed reflect a person's thoughts. This is why the eyes are anatomically closely related to the nervous system. The human eye is like a camera that perceives light from the outside world and forms images that are transmitted to the brain through the nerves. Light hits the outermost layer of the eye, the cornea, passes through the pupil and lens and becomes focused on the retina. The photoreceptor cells in the retina, including cone and rod cells, are the main cells in the eye that can sense light stimuli and transmit stimulus signals to neurons. These two types of cells resemble two little one-eyed people with only one leg. Although they both have only one leg, they have an unusually large number of toes and large feet, so they stand very still. The difference between the two cells is mainly reflected in their heads. The first one has a square face, wide chin, and pointed head, looking just like a cone. Therefore, its kind is called cone cells. The second has a round face with a high chef's hat on or a straight broom head, looking like hemp stalks from a distance. Its kind is hence called rod cells. Each human eye has over 7 million cone cells and over 100 million rod cells. The difference in the shape of the two determines the difference in their function. One of them can only feel the changes in light intensity, but cannot distinguish between the differences in color. Therefore, you need another kind of cell to help. If the latter is damaged or struck, it leads to the production of color weakness or color blindness.

It is important to emphasize that rod cells function with just a single photon of stimulus, producing signals that allow the eye to see objects even in dim light. The cone cells, on the other hand, need to be exposed to strong light in order to function and distinguish the colors of different objects. As a simple example, why can we only see white stars and not colored ones when looking up at the stars in the sky at night? The reason is that the light of naturally luminous stars has become very faint by the time it reaches the Earth after a long journey of several light-years or even tens of light-years, although the temperature of the surface of stars creates them with different colors such as red, yellow, blue, orange, and white. Our cone cells cannot identify it, so we can only see white color with the help of the rod cells alone. This gives us the white stars in the black night sky. The moon is much closer to us, so we see it sometimes as white, sometimes as red, and sometimes as orange.

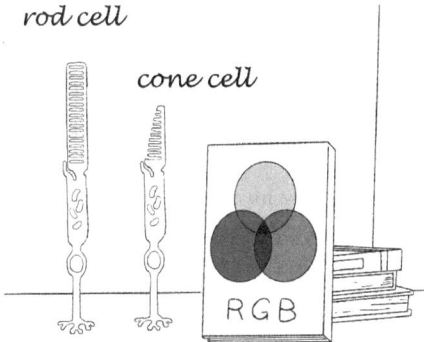

Why can cone cells see colors? This is because humans have three different types of cone cells in the retina that can be activated by three different colors of light: red light, green light, and blue light. As the three primary colors of light, their different combinations with changes in intensities together make up the colorful world we are able to see. However, for some people, either congenitally or later in life, the loss or damage of one of the three cone cells prevents them from recognizing one of the three primary colors. This results in what is often referred to as color blindness. This condition is not common in humans, but it is not rare either. You and I are surrounded by people who suffer from it. We, the human race, are very lucky animals in this respect because most animals only have one or two types of cone cells. Therefore, you could say that they're almost always color blind. The world is either black and white or monochrome—a bit of red, a bit of green, or a bit of blue. As with everything we see when we wear red, green, or blue glasses, what was once familiar immediately becomes strange and interesting at the same time.

The animal that needs to be mentioned here is the ox because, at the time of writing this book, the world was experiencing the ordeal of the COVID-19 pandemic. On the occasion of the New Year of the Ox in the Chinese lunar calendar, many buildings were decorated with large slogans: "It is not easy to be a Rat in 2020, but the Ox will turn the world around in 2021." However, oxen are quasi-color-blind because they lack green cone cells in their retinas. As a result, in the world of oxen, whether red and green or yellow and orange, it appears as the same shade of color. So why is the red flag waved to provoke the bull in Spain's world-famous bullfighting festival? You may not believe it, but the choice of red color is not for the bull to see. It is for the crowd to see, merely setting the mood of the festival. The bull does not care whether it is a red or a green flag. The reason it charged at the flag in anger was because the flag waving around annoyed it. In 1923, American George Stratton discovered through experiments that the color of the flag makes no difference in attracting the bull's attention. It's the amplitude of the wave that counts. The greater the amplitude, the more motivated the bull is to charge.

Since the absence of any one of the three types of cone cells leads to a less colorful perception of the world in humans and other animals, what happens if there is an extra type of cone cell? Theoretically, there are about 100 shades of color that each cone cell can distinguish. Three cells together can distinguish about 1 million colors, and four cells together can distinguish 100 million colors. Although the concept of four-color vision was first proposed by the Dutchman Henry Lucien de Vries in 1948, it wasn't until 2007 that the Englishman Gabriele Jordan discovered a person with four-color vision. If certain people have super powers, then people with four-color vision are definitely among them, as they can see more colors than ordinary people can.

The two most important systems in the human body are the nervous system and the cardiovascular system. As the engine of the animal body, the heart has to work round-the-clock to ensure that blood is constantly pumped to every organ and group of tissues in the body and nourishes every cell. The heart's beating is regulated by the nervous system, but it is mainly driven by its own cardiomyocytes. The cardiomyocytes, which are spindle-shaped, are born to be athletes. Even if they are separated from the heart, they can still beat there for a moment, one, two, three times … without stopping. Watching these pulsations, you might think you can hear the heart beating—thump, thump, thump. Cardiomyocytes have a "cousin"—myofibroblasts, which are longer and thinner in shape. In the middle, there is a very obvious cross section similar to that of bamboo. They are mainly present in the muscles of the body, being electrically stimulated by specific neuron conduction. They can also produce beating, but they do not move like cardiomyocytes. They only move occasionally and then stop, making them very lazy.

What is the secret of why cardiomyocytes and myofibroblasts can beat while other cells cannot? The main reason is that these two cell types are rich in two proteins that resemble ropes. Like a rope attached to an object, when the rope is pulled, the object can be pulled. When these two cells are stimulated by signals from the nerves, they promote a change in the concentration of calcium in the cells. This further causes the two rope-like proteins to pull on each other and eventually manifests itself in the movement of the cells. So even in the absence of nerve signals, a change in calcium concentration can cause the cells to beat, which is their own autonomous movement. In the case of the heart, if all the cardiomyocytes were dancing on their own, there would be a mess. That is why they are close together, and when a cell that is the leader beats, it quickly passes the beat on to the other cells so that they all beat to its rhythm. That is how the heart beats. Here we can try a small experiment. Find dozens of friends standing in a long line. One condition is that all the people are hand in hand. The other condition is that there are two or three meters between each person. Then, let the person on the far left say a whisper to the second person immediately next to him. The second person then passes the word to the third person, and so on, until the last person is on the right side. So that this last person speaks the whisper out, to see in which condition the phrase changes the least in the course of its transmission. The answer must be the first. For the muscles, if we lack calcium, it will cause the muscle fiber cells to twitch without our control, resulting in cramps. It's so sour that it's a real pain in the ass. Therefore, drinking milk or getting more sunshine to replenish the necessary calcium is essential for children to grow and for the elderly to prevent osteoporosis. It is also very beneficial for maintaining our heartbeat and preventing cramps.

The cardiovascular system, as the name suggests, is the heart and the blood vessels that connect the heart to the rest of the body. The main component of the blood that flows through these vessels is blood cells. Blood cells do not refer to a single type of cell, but to a group of cells in the blood that includes many types, at least a dozen. The bright red color of blood is mainly due to the presence of red blood cells, which are rich in hemoglobin, making the whole cell red. Additionally, the large number of red blood cells contributes to the red color as well. The shape of red blood cells is thick at the periphery and thin at the center, the opposite of the legendary UFOs that are swollen at the center and thinner at the periphery. If you happen to have a spherical piece of marshmallow or bread in your hand, squeeze the center for a while; the shape that emerges is that of a red blood cell. The other types of blood cells are basically colorless or white, which is why they are known collectively as white blood cells. Some of them like to stay in the lymph nodes or in the blood. The white blood cells that travel through the lymph vessels are round and small, almost the smallest cells in the body. They are called lymphocytes. The rest of the white blood cells are also basically irregularly round and slightly larger but look a little crooked, and some are freckled and look like they have a lot of sesame seeds stuck to their faces, explaining why most of them are also called granulocytes.

One of the most annoying things about being overweight is the stubborn fat that you cannot get rid of. It is mainly stored in fat cells. Because they are filled with so much fat, fat cells look like empty bubbles. However, they are not round and are a bit distorted, similar to when you stand in front of a distorting mirror and see that you have blown yourself up to an unbelievable degree. The bubbles are mostly white, some appear brown, and depending on the color they are called white adipocytes or brown adipocytes. If you accidentally pierce this bubble with a needle, the fat cells will be like a deflated ball. They will gradually

deflate and release a belly of fat floating in the water, just like oil droplets on the surface of soup when you drink it. So a person with a big belly also has big belly cells.

adipocyte

The most common reason why some people are fat and others thin is the more or less they eat, so where is all this food being absorbed by which cells? The epithelial cells of the intestinal villi in the intestine are extremely neat and tall columnar, being closely arranged together, forming a layer of dense membranous structure. Outside this layer of membrane, each cell grows long and thin bumps called villi, which are tightly packed together. If you don't look closely, it looks like the head of the Great Wall, with one bump and one depression after another, stretching endlessly. Their main function is to increase the surface area through the villi so that when the food digested by the stomach passes through here, it can maximize contact with the food, absorb as much of the nutrients as possible, and make them available to our body. The part that cannot be absorbed is excreted in the form of feces.

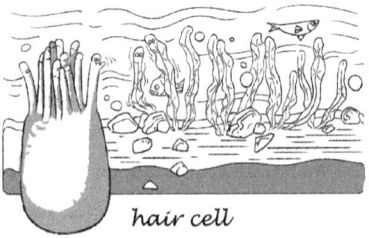

hair cell

In addition to the cells that have developed into villi in the intestine, there are two other places in the body where such cells exist: the ciliated cells in the airways and the hair cells in the ears. To deal with the dirt that enters the respiratory system and to prevent foreign bodies from entering the lungs, the mucus in the long respiratory tract sticks to these foreign bodies. Immediately afterward, little by little, under the oscillation of the ciliated cells, it pushes them into our mouths and nasal passages. We then spit them out in the form of phlegm or blow them out in the form of nasal snot, which is a disgusting waste secretion. When external sounds are picked up by our ears, the collected sound waves repeatedly hit the cilia of the hair cells. These cells convert the sound signals of different frequencies into electrical signals and transmit them to the nerves. The song "Seagrass Dance" is simply the most vivid description of this type of cell, and it's well worth listening to. Let's get a taste of seaweed through the lyrics of the song: "Like a seaweed, fluttering with the waves, dancing in the waves, regardless of the waves, I have my own fun. Float or sink, in the sea of people, with the waves changing. Oh, my life, such as a dream; where are you, my dear?"

Compared with all the cell types described above, the egg and sperm should be the two most familiar to the general public. The former is the reproductive cell produced by women, and the latter is produced by men. But there is a sharp contrast in size and shape between the two. The former is much larger than the latter and is round like an air-filled ball, while the latter is comparatively the size of a peg and similar in shape, with a small head and a long tail. It is interesting to note that the egg seems to be produced every time as an only child and seldom has brothers and sisters. If it does, it is at irregular intervals, not knowing at what time or exactly how many. Therefore, it gives the impression of always being unexpected and surprising, and one only has the opportunity to obtain it once a month. Sperm is completely different. It is not only produced every day but each time with the number of thousands. If we use the metaphor of siblings, it is perhaps a little inappropriate because the number is too great. Sperm production is like grabbing a handful of sesame seeds and then being scattered everywhere, and they cannot be counted.

In addition to the many types of cells found in our blood and organs, a certain number of cells are also found in the urine we produce every day. If more than 100 milliliters of fresh urine can be collected at one time, a few to dozens of cells can be harvested by centrifugation. Here is an explanation of what centrifugation is. It literally means to leave the center, and in the case of a circle, the center is the middle of it. A circle is formed when we take a string, tie a small object to one end of the string; then we take the other end and start throwing it as hard as we can. The harder we throw, the easier it is for the object tied to the string to be thrown out. We call this force away from the center centrifugal force. By using a special machine that produces centrifugal force, called a centrifuge, we can make the cells in a liquid come together and form a ball under the effect of centrifugal force. These cells are then expanded in vitro culture, and we soon see that there are many different shapes of cells. Some are cobblestone-shaped, some resemble a fried egg, and some take the shape of a pike. Recent studies have shown that they are a mixture of many cell types, but are dominated by renal epithelial cells. The kidneys play an important role as vital organs for water absorption and filtration in the body. Under constant water flow, a certain number of renal epithelial cells shed each day, circulate with body fluids, reach the bladder, and are eventually metabolized as a very insignificant component in the urine. This is like the banks of the river, under the influence of the water flow over the years. They always wash down some soil or gravel with the rolling water, either flowing to a distance or being deposited on the bottom of the water not far away.

Organelles of Cells

$\mathcal{L}$ike an apple, an animal cell is divided into three layers from the outside to the inside: cell membrane, cytoplasm, and nucleus, which correspond to the apple's skin, flesh, and core.

Just as a person has skin, a tree has bark, and an egg has a shell, every living thing always has its own protective layer to keep the outside world from harming its inside. The cell membrane is there to protect the cell. But this thin membrane-like structure is not as simple as you might think. Even with the most advanced optical microscope, it is difficult to observe the structure of the cell membrane. You can only see a line. As for the composition of the line, you have to rely on the electron microscope with better magnification ability. Under the electron microscope, scientists found that the film can be divided into two layers. Just as we buy things in the supermarket and are given a disposable plastic bag, we rub it vigorously twice only to find that it has a two-layer structure. Don't underestimate the process of rubbing twice. For a plastic bag, it may only take two or three seconds, but for a cell membrane, it takes two or three decades. As early as 1895, Charles Ernest Overton made tens of thousands of continuous attempts, far more than Edison's discovery of the filament. He conducted thousands of experiments and found that different chemical substances have varying abilities to penetrate cell membranes. The strongest ability to penetrate lipids was found in soluble substances. This led to the tentative conclusion that cell membranes might be composed of lipids. However, it was not until 1924 that Dutch scientists Evert Gorter and François Grendel discovered the automation of lipids in a monomolecular layer in the chemical solvent acetone. The spreading characteristics of dog, sheep, goat, rabbit, guinea pig, and human erythrocytes were placed in this solvent. By calculating the surface area of erythrocytes from different animal and human sources, it was found that the area of spreading lipids varied from 1:1.6 to 1:2.2, with the majority being 1:2. Although the volume and surface area of the erythrocytes varied from animal to animal, this ratio was essentially the same. They concluded that the cell membrane consisted of bilayer lipid molecules.

cell membrane

The early use of chemical analysis techniques has led to a further understanding of the composition of membranes, which are mainly bilayers of phospholipids. Each phospholipid contains a head and two legs, with the head facing outwards and the two legs facing each other. This divides it into an upper and a lower layer. Why is there such a strange arrangement? This is mainly due to the peculiar character of the phospholipid. Its head likes to drink water, while its two legs hate to get wet. So when they are put in water, they automatically gather together to form the above arrangement. Interestingly, some phospholipid molecules are like a playful child who likes to run around; some are like a ballet enthusiast who can stand on tiptoe and keep spinning; some are like a kung fu fan who likes to do somersaults from one layer of the membrane to another; and some are like a tap dancer who, once on stage, will shake their legs and feet involuntarily, playing beautiful music. Where do phospholipids come from? To put it bluntly, they are fats. Therefore, our daily intake of fat is necessary, at least in the synthesis of cell membranes. They are essential and beneficial.

Moreover, in the troubled sea of lipids, there are also numerous islands or boats made up of proteins, densely populated as far as the eye can see. They come in various sizes and shapes. As for the function of these proteins, which are mainly responsible for extracellular and intracellular transport or communication between the two worlds, sometimes they just send signals, and sometimes they can pass goods, so they are like telephone lines but also like oil pipelines. Here, I do not use the analogy of a boat for proteins by chance. The vast majority of proteins in the cell membrane are not fixed in one place but can navigate in a sea of lipids. As for how this phenomenon was discovered, it starts with a very clever experiment. The scientists labeled the proteins on the cell membrane of the whole cell with a fluorescent protein, as if the boat had been fitted with an electric light. Then, they shone a bright light on the left half of the cell, causing all the proteins on the left side of the cell to lose their fluorescence, just as if all the bulbs in the boat had been smashed. When night fell, people were surprised to find that the left part with no light started to slowly regain fluorescent proteins, as if the lighted boat was slowly sailing out of the bright water, illuminating the dark night, like stars looking from far away.

Passing through the cell membrane to the inside of the cell, the eye is greeted by a large and diverse world that is well organized in a way that no human world can match. A photograph of an overpass in Chongqing, China was once circulated on the Internet. In the most complex places, there may be six or seven layers with a dozen different lanes of traffic. If the microscopic traffic within a cell is magnified to the macroscopic world, there should be no less than thousands of layers or tens of thousands of lanes. This would be eye-opening! Conventional flyovers are really nothing compared to this, so you can imagine the complexity within the cell. Tiny cells maximize the use of their own resources in a limited space, often with one component carrying out several jobs. Take me as an example: at home, I am a father; outside, I am a driver; and in the institution, I am both a teacher and a technician. In addition to providing a major traffic channel for other components within the cytoplasm, the intracellular lattice also serves as scaffolding to prevent the collapse of the cell. Just like the steel skeleton of the National Stadium "Bird's Nest" and the skeleton of our human body, we have given this structure a visualizing name called the cytoskeleton.

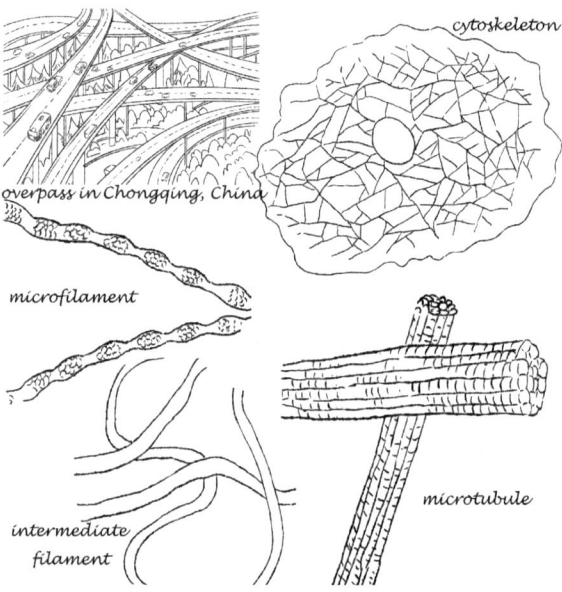

cytoskeleton

overpass in Chongqing, China

microfilament

intermediate filament

microtubule

As the saying goes, "All roads lead to Rome." We can take different approaches to achieve the same goal. The cytoskeleton has also adopted different strategies to serve the purpose of intracellular transport and support. Depending on the thickness of the skeleton, they are classified and named as microfilaments, microtubules, and intermediate fibers. The first are solid filaments made up of proteins, while the last two are hollow structures surrounded by different proteins. Whether it is a flyover or a bird's nest, once built, it is a permanent structure that will withstand sun, rain, wind, and storm. Unlike the real world, the cytoskeleton is always in a highly dynamic process of change: a cis-regulated microtubule there may have degraded and disappeared just in one second; at the same time, a microfilament may be synthesized from nowhere. These changes can occur in a fully automated state, but they can also be manually controlled. They are alive with the potential for a future high-tech artificial intelligence transport system. To achieve manual regulation, we can use two specific drugs: colchicine and phalloidin. At first glance, you wouldn't think these were made-up names from a magic show, but they are highly toxic substances found in the plant colchicine and the mushroom Amanita phalloides. Some seemingly beautiful flowers or wild mushrooms on the roadside often contain frighteningly deadly ingredients.

This densely packed cytoskeleton is only the tip of the iceberg in the cytoplasm. Its complexity far surpasses that of a snail shell used as a dojo, where the endoplasmic reticulum, Golgi body, lysosomes, and mitochondria are the real protagonists. We will refer to them collectively as organelles. With these organelles, the cell has a soul. As the saying goes, "The mountain does not need to be high; it is famous if there is a fairy. The water does not need to be deep; it has spiritual value if there is a Loong." We can also imagine that they are machines in an empty factory. With them, the factory can operate normally and produce products. In addition to these big guys, there are also some little guys, like the sorting robots that go back and forth in a courier company. These are ribosomes and centrosomes. Here, we describe what each of these big machines looks like, what they do, and who discovered them. We start with a heavyweight, Keith Roberts Porter, known as the father of cell biology.

Keith Roberts Porter

endoplasmic reticulum

Porter was born in Yarmouth, Nova Scotia, Canada, on June 11, 1912. He was the third oldest child in his family, with two older sisters. Like many, he graduated from high school at the age of 18 and went on to study biology at Acadia University, one of Canada's oldest universities. He had a wide range of interests, including being an accountant, trumpet player, assistant conductor of the school band, stage manager for plays, and president of the biology society in the student union. These experiences laid the foundation for the leadership skills he would develop throughout his life. After graduating from college at the age of 22, he went on to Harvard University, where he earned a master of arts degree and a PhD degree in biology, with his main academic interest in the developmental biology of frogs. After completing his degree, at the age of 26, he married a college classmate and moved to Princeton University to continue the research he had left unfinished. He invented a remarkable technique for separating the female chromosomes in fertilized frog eggs, leaving only the male chromosomes to form isolated haploid embryos. Only one year later, perhaps finding his previous research uninteresting, he returned to the Pathology and Bacteriology Laboratory at the Rockefeller Institute in New York, where he began to study tumors. Although the head of his new lab suggested he study what substances cause tumors, he was not interested and continued to play with his frogs while trying to understand the most basic biological question of what actually affects cell growth and differentiation. Unfortunately, less than two years into his new job, Porter and his wife had to stop their research as they were mistaken for tuberculosis patients and sent to a hospital for compulsory convalescence. However, he did not give up. With the help of a laboratory in the hospital, he accidentally discovered that changes in the culture environment significantly affected the morphology of Mycobacterium tuberculosis. Inspired, he set about establishing a set of standards for in vitro cell culture and eventually founded the National Society for Cell Biology.

Of course, the eventual establishment of these standards was fueled by another major event. In 1939, German physicists discovered that electrons could replace the role of light in conventional optical microscope imaging, greatly improving the resolution of the microscope from 0.2 micrometers to 0.2 nanometres, and built the first electron microscope. In 1944, Porter, along with other senior fellow apprentices and electron microscope engineers, made the first use of the electron microscope to observe cells after they had dried. As dried cells lose a lot of information, Porter improved cell culture and preservation conditions to better obtain true cell morphology. Finally, he succeeded in getting the first perfect super-resolution image of a cell. In this groundbreaking image, the center of the cell is thicker and electrons cannot penetrate it, so it appears as a black mass. Meanwhile, the periphery of the cell is thin, allowing the fine structure inside the cell to be clearly seen. They also discovered a new reticular structure, which they named the endoplasmic reticulum. Porter and many collaborators then used electron microscope to discover viral particles in chicken tumor cells and the microtubules in the cytoskeleton. Because electron microscope imaging is so demanding in terms of cell sample preparation, Porter waged a decades-long and tireless battle to get the perfect picture in terms of cell culture media, operating procedures, and tissue and cell sectioning. He founded the American Society for Cell Biology and trained many of today's cell biology researchers. In addition, because academic journals at the time were struggling to publish high-quality images and were unable to provide their readers with microscopic images of the highest quality, in 1954 Porter took the lead in founding an entirely new journal, the *Journal of Biophysical and Biochemical Cytology*. In 1962, it changed its name to the current prestigious academic journal, the *Journal of Cell Biology*.

So, what exactly is the endoplasmic reticulum Porter discovered? Normally, the first things that come to mind when we think of the components of an animal are lean meat and fatty meat. These are colloquial terms, and to put it in more technical terms, the main component of the former is protein and the main component of the latter is fat. How do cells, the smallest living units that make up the body, produce these proteins and fats? Thanks to the endoplasmic reticulum. It is a flat and multilayered mesh structure made up of membranes with some of them having many small raised particles attached to the surface of the reticulum, like the beef tripe we order when eating hot pot. Depending on whether these small particles exist, these reticular structures are called rough endoplasmic reticulum and smooth endoplasmic reticulum. Under the operation of this composite machine, amino acids that constitute proteins and lipid molecules that constitute fats are regularly spliced together by two different endoplasmic reticulums to form a variety of proteins and fats for use by cells and the organism.

After the production of protein products, they cannot go directly out of the factory immediately. They need to be further polished and labeled so as to form a qualified product that can be distributed to different places. To complete the above follow-up work, mainly rely on the Golgi body. In appearance, the Golgi body is very similar to a stack of naan from Xinjiang, China. It is slightly expanded at the periphery and has a concave and flat center. For a customized product that has just left the factory, it will be added with some additional features by the manufacturer according to the needs of different customers, then tagged with labels and text instructions, and finally transported out through logistics. Similarly, proteins then undergo various modifications in the Golgi body, such as a label of sugar, a chain of shiny phosphorus molecules, and so on. Once finished, these decorated proteins are transported to other organelles inside or outside the cell to participate in various cellular activities.

The curious will ask: why is it called the Golgi body? Why don't they call it the big cake body or the Xinjiang naan body, which are obviously very similar in shape? If I ask you one by one what the Golgi is, you will probably say that it is a person's name, right? In Chinese junior high school textbook, there is an

Golgi body

Camilo Golgi

EUROPA
CAMILLO
GOLGI
PREMIO
NOBEL
1906

article called "Seagulls," whose author is the Soviet writer Gorky. But we are talking about another Golgi here.* His full name was Camillo Golgi, and he was born on July 7, 1843, in Carthano, Brescia, Italy. It was a small village at the foot of a mountain. As his father was a doctor, Golgi studied medicine at the University of Pavia and graduated as a doctor at the age of 22. He soon became an assistant to a famous psychiatrist of the time and entered the field of brain research. Golgi then joined the Institute of Pathology, where he studied under Giulio Bizzozero, who was three years younger. Together, they made experimental medical observations with the microscope, thus formalizing his lifelong research on the nervous system and making him an established clinical physi-

---

* In Chinese, translations of Golgi and Gorky are the same (Gaoerji). The author here employed a form of wordplay, intentionally mistaking Golgi for Gorky, who is more well-known in China.—Trans.

opathologist by the age of 29. Later, at the instigation of his father, he moved to a small town hospital near Milan and applied for the post of chief physician.

With a small town hospital and limited access to scientific research, he turned the kitchen of his home into a modest laboratory where he carried out his life-changing experiment. From today's perspective, the discovery is trivial, but at the time it was groundbreaking. The whole field of neurology was trying to use the microscope to observe neural tissue, such as brain slices. However, there was a difficulty: the shape of transparent cells was impossible to observe. The result of existing staining methods for coloring was always unsatisfying, which led the whole field of research into a bottleneck. Golgi also began research in this direction, and on February 16, 1873, he wrote to his friend Niccolò Manfredi in his epigraph, "I have spent countless hours at the microscope and have finally discovered, to my surprise, that the use of silver nitrate instead of traditional dyes for staining can perfectly demonstrate the fibrous structure of the interstitial matrix in the cerebral cortex." This discovery laid the foundations for modern neurobiological research. Simple but extremely effective, it finally allowed people to look inside neurons and was successively called the "black reaction" and the "Golgi stain."

On the basis of this important discovery, Golgi returned to his former university successively as a professor and president. Everything just fell into place. Using the black reaction, he continued his observations of nerve tissue, discovering the presence of branches in the axons of neurons and the lack of fusion between dendrites. He also theorized that nerve impulses were transmitted by seamlessly fused axons. However, this theory was challenged by Santiago Ramón y Cajal of the University of Barcelona, Spain. Ramón y Cajal also used the black reaction but found that there were gaps between the axons of the neurons. Of course, later studies proved Golgi's theory wrong, but that did not stand in the way of his contribution to the development of neurology. He and Cajal were jointly awarded the Nobel Prize in Physiology or Medicine in 1906. In addition to the above, Golgi also made a discovery that was not a discovery. In 1897, while observing the staining of dorsal root ganglia after the black reaction, he saw an intracellular lattice structure that he named the "internal reticular organelle." Whether this structure was a separate organelle or not, and whether it had a function or not, could not be determined. The debate continued for nearly half a century until the advent of the electron microscope. At that time, the organelle was again observed and finally recognized and named Golgi body in honor of Golgi himself. In addition, in the 1990s, Italy issued postage stamps featuring Golgi's potrait and an image of a post-black reactive neuron in memorization of this great medical scientist.

However, there will always be a time when any machine or process makes a mistake, even if the probability of such a mistake is extremely low in the cell. What if something goes wrong and an inferior protein is produced? Don't worry, it's the lysosome's turn. It is like an intracellular dustbin, and it is a modern intelligent dustbin with degradation and sorting functions. In today's world of garbage sorting, the lysosome must be the founding master. Once a malfunctioning protein is thrown into the lysosome, it is degraded and turned from a protein into amino acids as the primary material. The released amino acids are then reused in the synthesis of new proteins, and the cycle continues. It is definitely a green and sustainable process.

The discovery of lysosomes was a happy accident. The story begins during the First World War. Christian de Duve was born on October 2, 1917, in Thames Ditton, a place near London, England, into a German-Belgian family. The family had to flee to England due to the First World War. After the war, when Duve was 3 years old, they returned to Antwerp, a port city in northwest Belgium where the inhabitants speak a dialect of Dutch while the official language is French. Growing up in such an environment, Duve learned four languages, laying the foundation for his later reading of scientific literature from various countries. The brief period of peace that followed allowed him to enter the University of Leuven at the age of 17, where he studied medicine. As for why he chose medicine, it was due to the same principle that most families follow today when guiding their children to choose their university major: to make it easier to find a job. With good grades, he was allowed to enter the laboratory to study and research the effect of insulin on sugar absorption with Professor Joseph Bouckaert in the school's physiology laboratory. In 1941, after graduating from university, Duve intended to give up his medical career and devote himself fully to studying the mechanism of insulin action. However, the Second World War broke out against his wishes. During the war, he served in the army for a time and was unfortunately captured by the enemy. Amid this misfortune, there was a silver lining—he managed to escape. Duve went back to school after the disaster and continued his unfinished work, spending four years analyzing the mechanism of insulin's action by biochemical means and writing a 400-page dissertation on the subject. Despite his successes, Duve was deeply concerned about his lack of knowledge in biochemistry. So, from 1946 to 1947, he traveled to the Nobel Medical Research Institute in Stockholm, Sweden, the Rockefeller Foundation at Washington University, and St. Louis, where he studied under four Nobel Prize winners. This one single year had deeply influenced a lifetime of his scientific research.

At the end of his traveling studies, Duve returned to his alma mater at the age of thirty and became a member of the faculty of medicine, teaching physiochemistry. He was promoted to professor four years later. During this time, he set up his own small independent laboratory with one technician and five students. This was a small but highly qualified group who were trying to understand the main enzymes that regulate sugar metabolism in the liver. However, as Duve himself said, his destiny seemed to be sealed. Although he did not achieve anything in his main field, he serendipitously discovered organelles that contained several enzymes. He had only observed these new organelles and did not know what they were about at the time. Back then, the whole field of cell biology had entered the era of microstructure through the use of the electron microscope. At the Rockefeller Institute in the US, Belgian Albert Claude and Romanian George Emil Palade, who were more than 20 years older than Duve, invented the differential centrifugation method and used it to obtain different organelles for use in electron microscopy. Differential centrifugation is based on the fact that objects of different sizes and weights sink at different centrifugal speeds and accumulate at different levels. This traps and enriches substances with similar properties. Using these new methods, the two men observed numerous new structures that were different from known organelles. However, they still knew nothing about their composition and function. Therefore, they urgently needed an expert in biochemistry to work with them. This led them to Claude's fellow countryman, Duve.

Christian de Duve

Albert Claude

George Emil Palade

Based on the close cooperation between the three of them, it was finally established in 1955 that the organelle containing multiple enzymes previously discovered by Duve was a new organelle with a completely new function. They named it "lysosome" and were jointly awarded the Nobel Prize in Physiology or Medicine in 1974. Duve also discovered another new organelle with a similar function to the lysosome but containing a different class of enzymes, the peroxisome. In addition to his scientific contributions, after winning the prize, Duve has also written a series of books on the theme of thinking about life, with a frequency of almost one every five years since 1984. These books are of great value, such as *Vital Dust: The Origin and Evolution of Life on Earth*.

Whether it is the activity of the organism, the activity of the cells, or the operation of the various organelles mentioned above, all need to be powered by energy. Just as the use of electrical appliances needs electricity and the movement of cars needs petrol, without energy, all movement can only tend to be static. We have developed a variety of technologies to obtain energy from nature, including wind power, solar power, thermal power, nuclear power, and so on. For animals, the main way to get energy is to eat. Of course, staying close to fire or sunbathing are also good ways. But for organisms, these can only be used as auxiliary methods. As for the food we eat, it consists of lean meat, fatty meat, vegetables, sugar, and so on. In more technical terms, it is protein, fat, fiber, and carbohydrates. So how are these substances finally converted into energy that can be used? It's thanks to the mitochondria.

Mitochondria are also specialized machines in the cytoplasm of the cell with a double-membrane structure. They have the property of being the generator of the cell and the energy factory of life. That is to say, mitochondria provide the cell with a constant supply of energy. For animals, the part that requires the most energy is the muscle fiber cells in the muscles. It is often said that "have no energy to run," meaning that the leg muscles lack energy. As early as the mid-19th century, scientists observed the existence of mitochondria in these cells, but only in a granular structure due to the microscopic technology available at the time. After half a century of technological development, German scientist Carl Benda looked again at these structures and found that these particles sometimes appeared as lines and sometimes as granules. Hence, the name "mitochondria" was given to them. In the century that followed, scientists from many countries worked tirelessly to unravel the mystery of mitochondria.

In order to better study mitochondria, we first had to figure out how to observe them better. Then, we isolated and purified them to facilitate research. This was achieved thanks to the contributions of several chemists, who used different staining methods and increased efficiency step by step before it became clear that one of the main functions of mitochondria is to release the energy contained in the food eaten into the body. In the biomedical field, there is a term to explain the process of releasing this energy, redox reactions, which is a combination of oxidation and reduction. So, what is oxidation and what is reduction? If we cut an apple, it doesn't take long for the color of the apple at the cut to change from white to brown. That's oxidation, mainly due to the oxygen in the air. If we put the sour orange juice on it, the color will be turned back to white. This is a reduction, mainly due to the vitamin C in oranges. In addition, the pigments we use in painting change color over time also due to redox reactions. So when chemists stained mitochondria with different colored dyes, they could guess what kind of chemical reactions were taking place in the mitochondria based on the color change. Of course, these studies were only the tip of the iceberg in the study of mitochondrial function. Combined with the rapidly advancing and sophisticated chemical technology of the time, a wide variety of chemical reactions and enzymes involved in mitochondria were gradually uncovered.

One of the leading figures in this field was the German chemist Otto Heinrich Warburg. Born on October 8, 1883, his family was of German-Jewish descent, members of a wealthy Jewish banking family dating back to the sixteenth century. Otto's father was a famous physicist of the time, and in the field of physics, there are Warburg coefficients and Warburg elements named after him. Otto had a good education and upbringing in such a family and grew up to study with the Nobel Prize-winning chemist Emil Fischer. He earned a doctorate in chemistry at the age of 23 and a doctorate in medicine at the age of 28. Building on his early learning experiences, he applied his knowledge of chemistry fully to his understanding of life processes. From the early uptake and use of CO_2 in plant cells to the later study of tumor cells, he utilized his expertise in chemistry to gain a deeper understanding of how life functions. He made two important discoveries in his life. The first discovery was of enzymes involved in redox reactions in mitochondria, which led to an understanding of how energy stored in food is transported by mitochondria. It was converted into electrons and hydrogen and released from the mitochondria, which earned him the Nobel Prize in Physiology or Medicine in 1931. The second discovery was the Warburg effect named after him, which not only secured him a place in history like his father but allowed him to surpass his predecessor.

mitochondria

Otto Heinrich Warburg

Paul Boyer

John Walker

molecular motor

So how is the released electricity and hydrogen in the mitochondria re-used? Is it possible that other organelles are really like cars that can run on electricity or hydrogen? The principle is basically the same, but the process is completely different. You have to start with the double-membrane structure of mitochondria. The outer membrane is shaped like a peanut shell with small holes all over it. The inner membrane undergoes a tumbling in volution, forming a labyrinth-like internal structure. Although not as complex as the

Minoan labyrinth that trapped the Minotaur in Greek mythology, the shape was very similar. Embedded in these inner membranes are countless precious and finely structured small motors. They are divided into two parts: one end is smaller and fixed to the membrane, and the other end is larger and rotatable. The electrons and hydrogen ions produced in the early stages form different voltages or concentrations, which make them underflow like water from high to low and push the motor to rotate as they pass through from one end to the other. Imagine this scene. Isn't it very similar to the production of hydroelectric power and the rotating water wheel? During motor rotation, degraded material is transformed into adenosine triphosphate (ATP), which contains a significant amount of energy and can be directly utilized. ATP, once synthesized in the mitochondria, is released through the pores of the outer membrane and transported to other organelles. There, it facilitates various chemical reactions and sustains life's movement. This story of waterflow was revealed like water flowing into the channel by two biochemists and took more than 20 years to complete. Initially, the hypothesis of intracellular water flow for electricity generation arose from the results of chemical experiments by the American Paul Boyer. Later on, it was proved by the Englishman John Walker by analyzing the specific parts that comprise the small motors. They jointly received the 1997 Nobel Prize in Chemistry due to this discovery.

Despite the structure and function of mitochondria being clarified, their origin remains an enigma. It is unexpected that mitochondria contain genetic material as independent organelles in the cytoplasm, unlike all other organelles except for the nucleus. As a result, researchers around the world have unleashed a wide range of imaginative ideas and finally proposed two reliable theories about the origins of mitochondria. The first theory suggests that mitochondria are part of the nucleus of the cell and are released through budding. The second theory proposes that some bacteria are ingested by the cell and, rather than being digested, they establish a symbiotic relationship with the cell. Eventually, they become mitochondria that serve the cell. Neither speculation has any substantial evidence but both seem to make sense. Who is right and who is wrong may be left for a future generation to solve the riddle of the century.

That's basically the story of the cytoplasm. Another important job for a tiny cell is to preserve genetic material. There is a tiny bit in the mitochondria, but it can only be thought of as a tiny shrimp stuck in the teeth of a shark. If the genetic material is damaged, it can lead to cell dysfunction in mild cases and various hereditary diseases in severe cases. Therefore, to be better protected, it is placed in the center of the cell and surrounded by a layer of membrane. As the core of the cell, this structure is called the nucleus. The membrane surrounding the nucleus is called the nuclear membrane. The structure of the nuclear membrane is completely different from that of the cell membrane in that it has many small holes on its surface to facilitate the movement of substances in and out of the nucleus. Now, we all know that the genetic material is mainly deoxyribonucleic acid (DNA), but if the DNA in each cell is stretched out, it can be up to two meters long. So, how does it fit into the nucleus, which is less than 6 microns in diameter? This can only be attributed to the wonders of nature. DNA will be the ultimate application of spatial geometry. It starts by wrapping around a few small proteins to form a polymer. Then, it winds like a pearl on a coil, forming a string of rosary-like structures. After this, these rosary-like structures fold into fibers and filaments. Finally, these filamentous structures superimpose on each other, forming advanced structures called chromatin. Imagine how a cocoon is formed with a single silk thread; you will have a good idea of the process. It's just that DNA involves many different patterns of regular folding, whereas cocoons are just a kind of messy, haphazard tangle.

Normally, this tightly coiled DNA material plays no part in the operation of cells; it just sits there quietly. However, when the need arises, this tight structure loosens in a very orderly fashion. The exposed strands of DNA are like a string of secret codes filled with instructions that guide the ubiquitous nucleotides in the cell nucleus to form a string of words that can be read by anyone with an eye for detail. We call these nucleotides, strung together in a string of clear textual meanings, ribonucleic acid (RNA). Once made, RNA will burrow into the cytoplasm through holes in the nuclear membrane and enter the endoplasmic reticulum with the help of ribosomes. This guides the synthesis and production of proteins. Since the DNA in the nucleus is not directly involved in the activities of the cytoplasm, the RNA that lies in between plays a role similar to that of a postman and is, therefore, also called a "messenger." Since this discovery was made half a century ago, I am afraid that if this type of RNA had been discovered today, it would have been called the "courier boy."

All the introductions above refer to the structure and activities of a single cell. In addition, a word you often hear is cell division, where one cell becomes two cells. What goes on? The introduction to cell division involves two very specific concepts: mitosis and meiosis. The former refers to the fact that before one cell becomes two cells, there is intense activity in the nucleus of the cell. The chromatin, which was already tightly coiled, first replicates itself, going from one to two, and then coils further to form rod-like structures called chromosomes. If they were a puddle of slime before, then they form shapely clumps of slime in this stage. Next, it's the turn of the cytoplasmic centrosomes, which

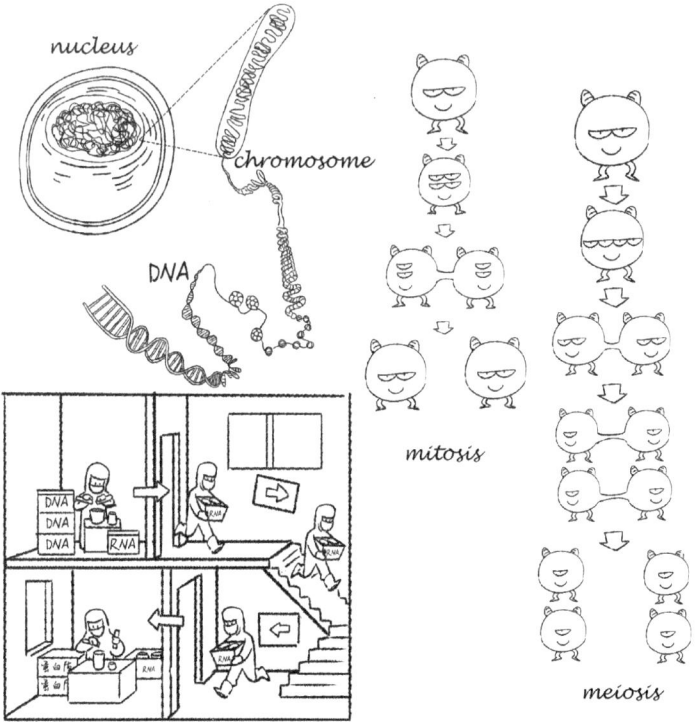

run to opposite poles of the cell. With the help of extended microtubules and microfilaments, they work in unison to pull and guide the cell toward the ends. Eventually, they split the cytoplasm and nucleus in two, forming two complete and identical cells. Repeated cycles of this process lead to an increase in the number of cells, a process we also call cell proliferation. The duration of each cycle usually ranges from a few hours to several tens of hours and thus determines the cellular growth rate.

In meiosis, it refers to the fact that chromatin does not self-replicate during the pre-divisional phase of cell division. Therefore, at the time of division, chromosomes are halved and distributed to the two cells of the progeny, reducing the amount of chromatin in each zygote to half that of the original cell. Hence comes the term "meiosis." In both male and female animals, none of the somatic cells undergo meiosis. It is only within the reproductive system that meiosis occurs, resulting in the production of reproductive cells that contain only half of the genetic material found in the cells of the rest of the body, which are the sperm and the egg. It is only when the sperm and egg meet and fuse to form a single cell that the normal number of chromosomes is restored.

The Social Circle of Cells

$\mathscr{T}$he introduction above is sufficient to show that the cell is a fully functional little fellow. The sparrow may be small, but all its vital organs are there. But it would be a mistake to think that cells are arrogant, self-absorbed, and unsociable. In fact, although there is a small part of the cell that is very shy and introverted, most cells are "sociable people" and even more, some cells can sacrifice themselves for friends.

In order to use the space in the body efficiently, all the cells are squeezed together almost seamlessly, showing the intimate relationship between each cell, as if they were glued together. It doesn't look like the cells are lined up neatly. They're not just back to back or face to face; almost every cell is hand in hand. And this kind of hand-holding can be described in various ways, some loose and some tight. Sometimes, it even involves passing each other goods through the act of hand-holding, which is a mutual exchange. Most of them rely on proteins embedded in the surface of the cell membrane, and these proteins are coated with another layer of sugar, which makes them stickier. Imagine spilling different concentrations of sugar water and wiping it with your finger. You can clearly feel the different levels of stickiness. Connections based on different patterns ensure that the cells are both neatly arranged and able to communicate with each other. When this arrangement changes, the cells can slip out of place, resulting in mixed but not necessarily bad results. Sometimes, the slipped cells cause damage; sometimes, the slipped cells go elsewhere to save the day. So, it is not possible to generalize.

Therefore, except for the cells in the blood that flow along the blood vessels throughout the body, the other cells are not all fixed in one place. The blood cells move around just as if a head-size coconut falls into the sea and flows down the waves to all corners of the world. Other cells, like sea turtles, shrink their heads and limbs when they are not doing anything. They either lie on the beach and sunbathe or stay in their nest and sleep. It is only when they are hungry do they barely move a few times. When they need to migrate, they travel further with the ocean currents. Whether in the body or in vitro culture, you will find that these cells are very naughty children. They not only like to run around but also have a special ability, such as shrinking bones. Once the cells encounter a channel smaller than their own head, they will shrink their body to drill through it and then restore their original shape. This is very magical. Especially for those square-headed guys, their body shapes would undergo obvious changes. First, the angles disappeared and turned into an oval shape, and then they turned into a shuttle shape, making it easy to travel through. How do they move? They mainly rely on the cell membrane protruding outward to form a tiny pseudopod. These pseudopods are very similar to the tentacles of an octopus but usually have more than eight legs to gradually move their fat body forward. In addition, they are very difficult to resist temptation. Once there's a substance that attracts them, they will desperately move toward these temptations until they can eat them. It's like a cat seeing salted fish or a dog smelling meat bones.

How can you prove that a cell can move its body that only has a round head or a big ass? We know that rats are very agile creatures, not only fast but also very clever. Only cats, snakes, eagles, and other animals can catch them easily. For us humans, if you try to catch them with your bare hands, the difficulty is like climbing up to the sky. Neurobiologists like to conduct experiments with mice, among which there is the water maze experiment. The researchers dropped a mouse into a pool of water, placed a platform in the water, and then observed how long it took the mouse to swim and where it swam to find a platform to rest on. They took a video and then used a computer to draw the mouse's swimming path. A normal mouse could easily find this life-saving platform, while some of the dumber mice had swimming paths that were a complete mess. Cell biologists use the same observation to record video of a cell being grown in vitro and then draw its trajectory. The trajectory of a cell is very similar to the trajectory of a mouse swimming in a water maze, either in a straight line, in a curve, or in a jumble. However, the former moves as slowly as a snail, while the latter moves like a rabbit. Someone once made a video of a bacterium placed next to a cell. The bacterium will run around involuntarily because of Brownian motion, and the cell will follow it until swallowing the bacterium into its stomach, which is very interesting.

For their own "right and left neighbors," cells can communicate with each other by talking to each other or moving their bodies a little. So, for those "distant relatives," what kind of communication happens between cells? Is it the use of ancient wolf smoke, telegrams, phone calls, or online communication? It is overthinking. These are unique means of communication for us humans. Cells do not have these skills. However, that's not going to stop them. In order to communicate with distant cells, they will use the way of spitting bubbles. The information to be transmitted is wrapped in the cell membrane, forming a closed bubble, and then let it float into the distance. Much like blowing bubbles, which we all loved to play with when we were young, the bubbles come in all sizes. With the wind blowing through them, some of them can float to very high and distant places. Once in a new place, these bubbles burst or fuse with the cell membranes of other cells, releasing the important information they contain. In a perfect long-distance message transfer, there is often more than one message in a bubble, perhaps several, dozens, or even hundreds to conserve resources. To distinguish it from other methods of communication, people have given these special bubbles a very stylish name—exosomes. These are the naughty cells that like to blow bubbles, and almost every one of them is a professional bubble blower.

Compared with the long history of organelles, whose discoveries can be traced back at least half a century, if not a couple of centuries earlier, exosomes are only less than 40 years old. In 1983, within a week, two different research groups reported the discovery of small bubbles with a diameter of about 50 microns in sheep reticulocytes. These were formally named exosomes in 1985 by Rose Mamelak Johnstone of Canada. Interestingly, the term "exosome" had been used four years earlier, but not to refer to these bubbles. Although exosomes were observed in blood cells in the early days, Johnstone and others discovered that they were also present in numerous other cell types and played important roles in communication. It's worth noting that the aforementioned figures who have contributed to the cellular field have been mostly male. Johnstone is one of the female scientists.

exosome

Rose Mamelak Johnstone

Born on May 14, 1928, into a Jewish family in Lodz, the third largest city in Poland, Johnstone's family fled to Montreal, Canada shortly after the outbreak of the Second World War. With the world in full post-war recession, life in Canada was even harder for working-class immigrants. Many of them could only afford to send their first child to school, while younger siblings had to drop out of school to help the family. Despite being the second of four children, Johnstone was able to complete high school and attend McGill University to study microbiology thanks to her mother as an active feminist. During her time at university, she worked hard, won several scholarships, changed her major once, and eventually majored in biochemistry. She graduated at 22 and earned her PhD at 25. Since then, she has been involved in research as an assistant professor, associate professor, and professor. Her working life was not smooth as she was deeply affected by the sexism of the time. She fought against it vigorously until she retired at the age of 67. Throughout her life, she did not only achieve remarkable research results in the scientific field but also contribute greatly to the advancement of women's position in academia.

Of course, not all information has to be in the form of bubbles in order to be transmitted. For relatively stable information substances, the cell will simply spit them out and let them travel outside the cell on their own. These information substances can either increase the activity of other cells or cause other cells to move into position. Each category is a large family with many family members. These families include growth factors, cytokines, and hormones. Both in the past and today, some marketers have promoted the idea that cytokines can enhance skin beauty, but what they're actually referring to is a concentrated form of these cell-secreted factors. Different types of cell-secreted factors often have big differences, with even larger difference divergences between their concentrations. How to ensure that the concentration reaches an effective level is a key issue. If the concentration is too low, it often fails to play a sufficient stimulating or inducing role; if the concentration is too high, too strong stimulation and induction will be counterproductive, leading to a malignant transformation of cell fate. Therefore, the use of cell secretion factors is definitely a double-edged sword. In addition, there is another key class of substances involved in cell-secreted factors, namely hormones. While cytokines are mainly proteinaceous substances, hormones are mostly classified as lipid substances, and their discovery and use predate cytokines by a wide margin. The best-known cell-secreted hormones are growth hormone and adrenaline. The former is associated with a child's height, while the latter is responsible for the surge of blood during emotional excitement.

In addition to proteins and lipids, there is another group of substances directly secreted by cells that are also very robust: DNA. The research on this has ushered in the era of non-invasive DNA prenatal testing, which has benefited tens of thousands of pregnant women worldwide. The man who opened the door to this era is Chinese scientist Yuk Ming Lo, also known as the father of non-invasive DNA prenatal testing. As for how he invented this technology, it all started with a dinner party, a bowl of instant noodles, and a film to talk about.

Hongkong, China

Yuk Ming Lo

Non-invasive
Prenatal Testing

On October 12, 1963, in Hong Kong, China, Yuk Ming Lo was born into a family of psychiatrists and music teachers. Benefiting from the good family environment and parental upbringing, he studied at Cambridge University and Oxford University in the UK successively, where he spent ten years obtaining his bachelor's, master's, and PhD degrees in Literature, Medicine, Surgery, and Philosophy. It was during this period that he was first introduced to the then cutting-edge technology of polymerase chain reaction, or PCR for short. This technology can amplify a small amount of nucleic acid into large quantities in vitro, just like a photocopier. Learning the technology was not an end in itself, but a means to an end. Yuk Ming Lo, after learning and mastering the PCR technique, thought about the possibility of using the technology to detect a fetus inside a pregnant woman. Why did he want to do this? During his internship at the Department of Obstetrics and, he noticed that prenatal testing of pregnant women was mainly done by amniocentesis, a technique that carries certain risks of injuring the fetus. This made him wonder if there was a better way to test. During a dinner with his classmates, they discussed a topic that all young couples tend to consider—whether they want to have children or not and whether they prefer a boy or a girl. Lo thought long and hard about this, and then he got a sudden flush of inspiration: If the fetus is a boy, its chromosomes will be completely different from the mother's, and it should be possible to detect this by taking cells from a pregnant woman's blood and using PCR technology to detect chromosome fragments in them. No one had ever done this test before, so Lo took this exciting idea and ran with it. He used PCR to test 19 pregnant women, and in 12 of them, he detected cells that were male. Finally, all of these 12 women indeed gave birth to boys, whereas those who didn't have such cells gave birth to girls. The discovery was eventually published in the leading clinical research journal, *The Lancet*, in 1989.

In 1997, after the handover of Hong Kong, Lo returned to China after completing his studies. He wanted to continue his research on prenatal testing, but due to the small number of fetal cells in the blood of pregnant women, test results were often inaccurate. False-positive or false-negative results could be seen from time to time, leaving him at a loss. Not until he read an article in the journal *Nature Medicine* about the detection of tumor DNA in the blood of cancer patients was Lo profoundly inspired. Since the DNA of tumor cells can be detected, there is no reason why the DNA of a three- or four-pound fetus should not be detected as well. With a hint of excitement, he started to detect fetal DNA in pregnant women. But this time he was not so lucky because the pregnant woman's plasma was full of too many proteins and her own DNA. The limitations of the technology at the time made it difficult for him to extract a small amount of fetal DNA. However, as the old saying goes, "The god of fortune always favors those who are well prepared." This time, as he was boiling instant noodles, he suddenly had the idea that if he took the blood of a pregnant woman and boiled it, what would happen? It was a wild idea, but he did practice it. The result was surprisingly good: the fetal DNA was condensed 100,000 times. He published this discovery in the prestigious medical journal mentioned before again, reporting for the first time that free fetal DNA could be found and successfully detected in pregnant women. Eight years of perseverance, from the previous dinner to this bowl of noodles, rewrote the history of prenatal testing and changed his life. All sorts of honors have come since then.

The honor didn't stop him from thinking and moving forward. It just made him understand that "the greater the ability, the greater the responsibility." In prenatal diagnosis, one of the most important screening tests is Down syndrome screening. And because the screening cannot be only qualitative, it must also be quantitative at the same time, a higher test accuracy is required. It is a difficult challenge to find a better way to use the technology that has been established in the past to screen for Down syndrome and other hereditary diseases. To this end, he has been experimenting with more accurate digital PCR technology and technology that can directly read out the exact sequence of DNA, gradually increasing the accuracy of screening for Down syndrome to almost 100%. After that, there was a bigger goal that appealed to him—screening for other genetic disorders. This would require knowledge of the DNA sequence of the entire fetus, making it an almost impossible task. It wasn't until the summer of 2009, while he and his wife were watching *Harry Potter* in the cinema, that the initial H in the opening credits slowly appeared on the screen and struck him like a bolt of lightning. The shape of the H resembled the structure of a chromosome, and he immediately thought that half of a child's genetic material comes from the father and half from the mother. If he had the parents' DNA data, he would be able to splice together pieces of the fetus's DNA. In 2010, he and his team finally obtained the entire DNA profile of a fetus from the blood of a pregnant woman, taking another solid step toward comprehensive non-invasive prenatal diagnosis. At present, this technology has completely transformed into a huge wheel of social progress, carrying humanity into a better future in both the industrial and medical fields.

After listening to Lo's story about cell secretion detection, I'm sure you must be excited! It's the power of technology to benefit humanity, isn't it? But not all stories are so inspiring, and the next story to be told will surely make you sigh.

Anyone who has ever been to the hospital for a medical check-up or visit will have experienced that, no matter how serious the illness is, the doctors usually advise us to have a few tubes of blood taken and wait for the test results before deciding on the next step in treatment. As you can see, testing for substances secreted by cells into the blood has become an essential part of the process of seeing a doctor. However, the current testing process has many shortcomings. First, a relatively large amount of blood is taken from the patient; second, the waiting time for test results is relatively long; and third, only a few indicators can be tested once. It would be remarkable if there were a way to solve all these problems at once. A Silicon Valley-based company founded in 2003 called Theranos was working to solve these problems. The company, whose name is a combination of "therapy" and "diagnosis," had a vision of taking a drop of blood from a patient's fingertip—literally a drop of blood, typically 50 microliters—and placing it in a plastic box with a size of a credit card and analyzing for more than 200 tests in a short period of time within a machine with the size of a microwave oven and. Whether it's routine blood tests, liver function, carcinoembryonic antigen, or anything else, it wouldn't matter. Were this project actually realized, it would not only revolutionize testing technology but also greatly aid treatment. It could definitely rewrite the existing diagnosis and treatment system. As a result, within just a few years of its establishment, the company had gained the favor, support, and participation of many prominent political and business figures in the United States as a veritable all-star lineup. The company had received more than $600 million in investment and was valued at more than $9 billion, despite not having any sales. It had also formed a strategic partnership with Walgreens, one of the country's largest drugstore chains, so that anyone who wants a blood test no longer has to go to a hospital, but can simply go to a drugstore and get the results as soon as they arrive. Just when everything was going in the right direction, the building came crashing down. The colorful soap bubbles in the sunlight faded, burst in the air, turned into fog flowers, and disappeared without a trace.

It was like rollercoastering, and Elizabeth Anne Holmes was in the car maneuvering it all. One of the questions children are often asked is what they want to be when they grow up. Many answer either a teacher or a doctor, or they just don't know. Little Holmes, however, knew exactly what she wanted to be—a billionaire. That's why, at the age of 19, she dropped out of college after just one year with a patent for a premature invention and founded the aforementioned company, Theranos. With a mother from one of the wealthiest families in the United States and a father who graduated from the prestigious US Military Academy at West Point and ran a good business, she was blessed with a life in the rich world and connections that made it easy for her to get her first investment from a neighbor. Later, with the help of her father's connections, she was able to promote the immature product to the military for use and thus gain government approval. If it were up to external factors alone, it would still be difficult for Theranos to take its place in Silicon Valley, where start-ups are flourishing. As the company's founder, young Holmes has her own strengths. She took the company's 360° dead-angle monitoring and all-around control, whether it was the corridor, living area, or R&D area, as well as the computer network. As long as there were any employees or outsiders questioning the company's products, even colleagues discussing the product, they would be immediately expelled. They would also receive an email personally sent by her, which was copied to all staff. The content of the email was to pack one's things and leave, as well as a confidentiality agreement. To the outside world, it all looks like a commercial company's way of keeping technology secret. However, in reality, it's merely a gilded façade, shabby and decayed beneath the surface, constantly threatened by the fear of being exposed. In addition, Holmes has shown social skills beyond her years, with a strategy of covering up one lie with another, both for investors and partner companies. With so many celebrities backing her, people firmly believed in her lies; even if a few questioned it, they were too afraid to say anything. And when lying becomes a habit, even she herself may be convinced. As the false comes true, the true can also become false. However, as a product that directly serves patients, if it is adulterated, the loss of money is a small matter, human life is much important. When the product is applied in real-life situations, unreliable test reports and exaggerated results will not extinguish the anger of frontline medical institutions and patients in the end.

Although there are differences in the substances that are connected and exchanged between most cells, they are basically regular items, as if they were delivered by a courier. But there are some very specific cells that don't do any of the above. They just concentrate on their own job of delivering a certain type of substance or information with a high degree of efficiency, equivalent to special military transport. This is the neuron of the nervous system. Due to the dramatic changes in the shape of neurons, it is no longer a round cell in the traditional sense, nor a square or pentagonal cell, but a body with multiple long tentacles. Each tentacle is topped by multiple thin branches, forming a localized extension where the tiny branches of different neurons are in proximity to each other. This is similar to two disks leaning against each other, adjacent but not touching. Because of this fine structure, when neurons receive external signals and stimuli, they rapidly transfer substances—mainly neurotransmitters and electrical currents—from one disk to the next. This speed of transmission is so fast that it can be considered a masterstroke in the body, enabling us to react quickly.

As mentioned earlier, these various substances that promote intercommunication and communication between cells all play their own positive role under normal circumstances and are tightly regulated by the cells, both in terms of when and where they are secreted and how much they are secreted. When these factors are disrupted, the result can be inflammation in the mildest cases or a severe inflammatory storm in the most severe cases. Those who have just experienced the new COVID-19 pandemic will not be unfamiliar with the latter, as one of the causes of patient deaths often mentioned in media reports is the famous cytokine storm. As a result of the rapid production of large amounts of cytokines in a short period of time, other cells are unable to process them properly. This leads to cellular dysfunction and failure of the organs in which they are located, ultimately resulting in the death of the organism. In addition, in the late stages of immunotherapy for tumors, one of the main causes of patient death is also the cytokine storm. Like a hot summer's day, there is no need to mention the comfort of gentle wind and rain. However, when the wind and rain suddenly escalate into a storm, everything becomes purgatory.

If you ask me whether there are more cells or secretions in the tissues of the body, I can't answer you. However, if you were to ask me, assuming all those cells were gone, would those factors still be present and functioning? My answer would be yes. Cell secretions contain many types of substances, including proteins, lipids, and polysaccharides. When mixed together, they tend to interact and cross-link with each other, providing a three-dimensional scaffolding structure for these cells. When the cells are completely removed using a special perfusion technique, a fully decellularized scaffold is obtained. These scaffolds retain abundant cellular secretions, so that when the cells are reinfused into them, they can easily form a new tissue or organ as if they had returned to their old home. They can then be used for transplantation therapy. In addition, based on the mimicry of these natural structures, scientists have been able to construct these scaffolds directly in vitro using synthetic or 3D printing techniques. Cellular tissue engineering has been driven by the variety of scaffold materials that can be used to mimic cellular secretions by adding active factors. The best-known of these scaffolds is hydrogel, which accounts for almost half of tissue engineering today. It is a jelly-like substance in which chemical molecules of different lengths and sizes are linked together by chemical cross-linking techniques to provide structural toughness and the ability to attract water, which is essential for sustaining life. Scientific studies have shown that growing different types of cells on the basis of cytokine-containing hydrogels has shown great promise in repairing nerves after spinal cord injury and in restoring joints after injuries that are virtually incurable with conventional medical treatments.

With a better understanding of cellular secretions, doctors have also used a variety of cytokines to treat diseases and harness these factors for human benefit. By understanding the functions of the different factors and the cellular functions they affect, we have been able to target individual cytokines to treat specific diseases. In addition, there is a large class of secreted proteins that should not be overlooked: antibodies that fight off invading enemies. Whether they are bacteria, viruses, or other "bad" cells, the cells in our body can recognize them quickly and accurately. From there, antibodies are secreted to fight these invaders. Antibodies are mainly produced in blood cells, so once produced, they can quickly travel through the blood vessels to the site of invasion and exert their therapeutic effect. As an antibody therapy that starts early but catches up late, its real spring has only just begun. As the production technology matures, the process becomes more stable and the cost decreases. It is believed that this cutting-edge biological therapy will definitely shine in the near future for the benefit of ordinary people.

Another example of intercellular communication that is well-known but not yet understood is the communication between the cells of our gut and the cells of our brain. When we are very anxious, we often feel sick in our guts.

When our guts feel sick, this in turn affects our thinking. There is growing evidence that this is the result of an interaction between the two through cellular secretions that travel through a long cycle of bodily fluids. For the latter in particular, the term "gut feeling" is used very vividly to express "intuition."

gut feeling

Every living thing has a life cycle, and the cell is no exception. After fulfilling their mission, many cells will voluntarily leave the stage to allow new cells to emerge and take over. They leave in a variety of ways, including necrosis, apoptosis, autophagy, and so on. As seniors may have a sunset glow, it seems to be the same for cells. In many cases, the deceased cells do not just disappear but also show strong affection and nourish the neighboring cells. This is illustrated in the case of "the falling flowers are not unfeeling things, but turning into spring mud and protecting their descendants."

The three ways of leaving mentioned just now mean different histories and different endings for the cell.

Let's start with the most common form of death, cellular necrosis. If you compare it with the way of human death, cell necrosis is more like being killed, and mostly involuntarily, such as with knives and swords in the cold weapon era, with guns and cannons in wartime, and car accidents in peacetime. The result is often a bloody and ugly death. For the cell, when it encounters knife cuts, fire blazes, or the light irradiation of the Japanese anime character Ultraman, the first thing that happens is that the cell membrane ruptures. This is followed by a belly full of parts popping out as uncontrolled as Doraemon the Explorer's treasure chest-like pockets. Everything comes in a sudden. For the surrounding cells, this is no pie in the sky. They are really not blessed to handle the situation, and a series of adverse reactions will ensue.

The second type of death is apoptosis, also known as programmed cell death. From the literal meaning of the word, it is easy to guess that cells follow a program step by step toward extinction, similar to a computer program that has been set up so that once activated, it can be completed step by step. This kind of death is like the end of human life. Compared with the bloody storm of cell necrosis, this active cell death is much gentler. Although the cell membranes do not rupture at this time, they will invaginate or bud out in different parts, resulting in a complete cell being divided into a large number of different sizes. These smaller cells are wrapped by the cell membrane into a number of small independent bubbles. Each bubble is wrapped in a different part to prevent the contents from leaking out. We know that tadpoles are born with tails, and when they finally turn into frogs, their tails disappear gradually rather than breaking off all at once. Why does it disappear? The main reason is that the cells in the tail undergo apoptosis. Programmed cell death is, therefore, essential for the normal development of the organism, not only because it favors the growth of organs but also because it plays an irreplaceable role in the cellular turnover that takes place every moment.

apoptosis

John Foxton Ross Kerr

The discovery of apoptosis is largely attributed to John Foxton Ross Kerr, who, as a PhD student in the UK in 1962, was given a project by his supervisor to study the cellular processes that occur in the liver when blood vessels in the liver are ligated and the liver tissue is crushed. In three years during the project, he found that although some of the cells were not necrotic, they fell apart and eventually disappeared. However, their membranes did not rupture. This phenomenon was similar to necrosis but not exactly the same, so he called it necrotic crumpling. After graduating, he returned to his native Australia and the University of Queensland, where he continued to observe this new type of cell death using electron microscopy and made detailed notes on vesicle formation. In 1970, Robert Currie, Head of the Department of Pathology at the University of Aberdeen in the UK, came to Brisbane on a short exchange visit. When Kerr showed Currie his electron microscopy results, Currie was immediately interested and told Kerr that he and Andrew Wyllie had been working together on the same subject. Wyllie had observed a similar phenomenon in kidney tissue. At Currie's suggestion, Kerr came to the University of Aberdeen to work with Wyllie to study the phenomenon in more detail, in different cell types and under different conditions. In 1972, the three co-signed a paper giving the phenomenon a new name, apoptosis. The name apoptosis was suggested by a professor of Greek at the University of Aberdeen. In Greek, apoptosis means the fading of petals from a flower or leaves falling from a tree, which was a very vivid description of the phenomenon they were observing. In the early years, apoptosis did not attract the attention of experts. However, as research deepened, it became one of the mainstream directions in cell research. The famous scholar, Professor Wang Xiaodong, made outstanding contributions in this field, found a series of keys to controlling cell survival and death, and was elected a member of the National Academy of Sciences of the United States of America. He is the first person from mainland China to receive this honor after the reform and opening-up.

As early as the mid-1990s, Duve, the discoverer of lysosomes, identified the phenomenon of cellular self-phagocytosis and named it cellular autophagy. However, it didn't attract much attention from researchers until a laid-back

Wang Xiaodong

style Japanese scientist, Ohsumi Yoshinori, finally and inadvertently catapulted autophagy from the back burner to the front burner. Ohsumi Yoshinori was born on February 9, 1945, in Fukuoka, Japan, during the Second World War. Shortly after his birth, Japan announced its surrender, leaving the country in dire straits. Due to his family's poverty and lack of food, he was weak from an early age with no particular strength in the aspects of sport, art, and literature. But fortunately, the vast natural surroundings allowed him to play happily in the fields, streams, and mountains. Catching fish and birds was his daily routine. In addition, his older brother, who was at university, would bring him many popular science books from time to time. This allowed him to pass the time while broadening his horizons and developing an interest in science. In high school, Ohsumi Yoshinori was immediately fascinated by the various strange chemical reactions he saw and decided to study chemistry when he entered the University of Tokyo. However, he soon found chemistry boring and switched to the fledgling modern molecular biology program. During his graduate studies, Japan was going through a politically turbulent time with various social movements taking place, especially in the Tokyo area. To better focus on his research, he took the plunge and moved from Tokyo to Kyoto University during the second year of his PhD program. There, he met and married his wife at the age of 26. After graduation, he left Japan for the first time to do postdoctoral research at Rockefeller University in the US on the advice and recommendation of his mentor due to the difficulty of finding a job. However, the direction of his mentor's research was embryonic development, which was very different from the research he had done on E. coli during his PhD. In his studies, he could not get the hang of it, and the progress of his experiments was slow, causing him a lot of headaches. At the time, a master of "playing with" yeast came to the lab, so Ohsumi Yoshinori changed the direction of his research and started to play with yeast under this master's guidance. After that, instead of crushing grain and throwing it away like a monkey crushing corn, he immersed himself in the world of yeast for the rest of his life.

At the age of 32, Ohsumi Yoshinori finished his postdoctoral work and finally got the chance to return to his home country to work as a research assistant at the University of Tokyo. Although specialized in E. coli research, his supervisor in the lab is very nice, so Ohsumi Yoshinori was able to continue playing with his yeast there. In the meantime, he began to work on the study of vacuoles in yeast, which were recognized as a rubbish bin among cells by the mainstream, with nobody being interested in them. Ten years passed, and at the age of 43, he was fortunate enough to be offered a faculty position at the University of Tokyo, where he finally established his own independent research group. In the beginning, the laboratory conditions were very simple, with only a vibrometer, an incubator, a spectrophotometer, and an optical microscope. It was with these instruments, and the concerted efforts of a few graduate students, that he discovered the key gene that controls the production of vacuoles in yeast, and went on to discover dozens of related genes and the functions of these genes, which brought the long-standing superficial observation of the phenomenon of autophagy to a deeper level and led to him being awarded the Nobel Prize in Physiology or Medicine at the age of 71. Looking back on these discoveries, he attributes them to two things: following his own curiosity rather than doing what's hot and having luck. As the vacuoles in yeast are large enough and like to move around, he is allowed to observe them with a rudimentary optical microscope; if he had to use an electron microscope to do his work, he would be out of the game. Today, Ohsumi Yoshinori is still working on autophagy, but the focus of his research has shifted from yeast to animal cells, in the hope of linking autophagy to more diseases and benefiting humanity sooner rather than later.

Cells Can Be "Nurtured" Too!

$\mathcal{W}$hen it comes to the word "nurture," it can be difficult for those not involved in biomedical research to understand: how should cells be nurtured? Is it like raising a child or a small animal, giving it water, food, and clothing every day? If you think so, you are only half right because compared with children, cells are really too delicate, and if you are not careful enough, they will lose their temper, go on a hunger strike, or even die. So, how exactly do you raise cells?

First of all, we need to create a very clean room. Is it enough that there is no dust or confetti on the floor and that it looks neat and tidy? This can only be seen as a first step because cells are most afraid of the little invisible things like bacteria and viruses, especially bacteria. To reduce airborne bacteria to a safe level, the air entering the room must be filtered extremely efficiently. Since you can't see bacteria, how can you test whether the room is clean? It is necessary to borrow knowledge from microbiology. Find a vial filled with a high-temperature sterilization medium, and then place the vial in different places in the room. After a few days, see if there are thriving bacterial plaques in the vial. According to the size of the room space and the number of plaques, calculate the ratio between the two. Using these ratio parameters, along with other suspended particle counts, it is possible to quantitatively analyze how clean this room is. Typically, the minimum level required by the cell is in the millions, and it would be even better if it could reach the 100,000 or 10,000 level. Now you know, just the air requires so much care. Is it enough just to filter the air? There is still one step to go. The usual air filtration is not so efficient, and to further improve cleanliness and reduce the number of bacterial particles, it is also necessary to add ultraviolet light. This is like a natural enemy for bacteria, as it can kill almost 100% of them. Of course, ultraviolet light can only kill bacteria on the surface of objects or in the air that it directly illuminates. It is less effective against bad elements hiding in dark corners. For this reason, ultraviolet light also has a special ability: it can break down and reassemble the ubiquitous oxygen in the air into the fishy odor of O_3. Don't underestimate O_3. It is a natural killer of bacteria and viruses. Compared to ultraviolet light, it can get into all sorts of nooks and crannies, pull out the bad guys, and beat them hard. And all that is just a prelude to raising cells well.

Now it's our turn. If you rush straight into the cell culture room, however, you could be in big trouble. As if you were developing films in a dark room, suddenly opening the door may not destroy everything, but at least destroy the work that has been done before. Just as a chef needs to wear a toque to fry, a doctor needs to wear a white coat to see a patient, and a beekeeper needs to wear protective gear, each profession has its own specific uniforms. As a cell keeper, this uniform can be considered fully armed, including disposable hats, disposable masks, disposable gloves, back-to-front clothing, and shoe covers. All this is not just for the sake of looking good, but for self-protection. It helps to prevent contamination or entry of toxic reagents or viruses and to prevent contamination of the cells. Although the air is already clean, the coats we wear, the air we exhale, and our hair harbor a large number of bacteria and viruses. If we are not careful, these can drift out on the air currents and create a new source of contamination. To minimize the risk, the above equipment is essential. After putting on the equipment, you need to spray some 70% alcohol on your gloves and then go into a room called the air shower for the air shower and stay there for a while. The reason you need 70% alcohol is that this is the strongest concentration for killing bacteria and viruses. Lower or higher concentrations are not optimal or even ineffective. And what is an air shower? We know that when we take a shower, the water comes down from above. An air shower is a way of using the wind instead of water, blowing from above and below, with the aim of blowing as much dust and fine dirt off the outside of your clothes as possible. Once you have completed the above steps, you can officially begin the process of culturing cells.

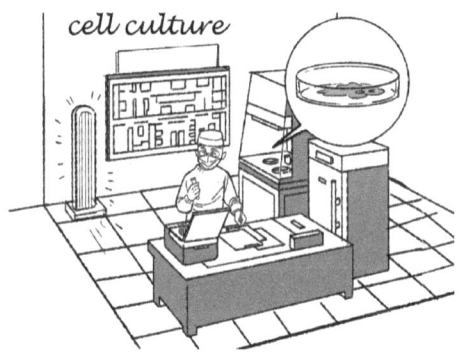

cell culture

Our normal body temperature is 37°C. A temperature too high or too low is uncomfortable, and so for the cells in our body. Therefore, we have prepared a box that can maintain 37°C. At the same time, in order to better simulate the comfortable environment of the cells in the body, we also need to add a certain amount of humidity and oxygen. For the former, we simply need to place a small pot of water in the bottom of the box and let it evaporate naturally. For oxygenation, we need to prepare a large bottle that is often seen in hospitals. However, it is not filled with oxygen but with liquid CO_2. The researchers introduce the gas into the box and then precisely control the proportion of oxygen by adjusting the size of the inlet, which is usually 15%. Do all cells need this temperature and concentration of oxygen? For most mammalian cells, yes, but insect cells are nurtured at around 30°C. Some specialized cell types also need lower oxygen concentrations, even as low as 5% or 1%, a condition we call hypoxia. Although it is extremely difficult for our bodies to live in a hypoxic or low-oxygen environment, it is paradise for some cells. As we can see, there are both similarities and differences between cells and the human body. No generalization can be made.

With this complete setup of the box above, isn't it safe to put the cells in? Basically, it's OK. But just as you can't eat out of a big pot of rice, you have to serve it in a smaller bowl, right? The same is true for culturing cells. In order to better distinguish between different types of cells and prevent them from being mixed together, we need to put the cells in a small bottle. However, the lid cannot be too tight. This is not only to ensure that oxygen and steam can flow freely but also to prevent the cells from being directly exposed to the air. In such a bottle, add a nutrient solution that the cells like to eat and drink. The cells can grow quickly. After decades of hard work, bit by bit, through experiments, scientists have figured out ingredients and methods to produce different kinds of nutrient solutions. Different varieties of nutrient solutions with a more professional name called culture medium can be directly bought in today's market, just like the wide range of beverages on shelves in the supermarket. Do not underestimate the cells. They are alive, so they need to eat and drink at the same time. However, they will also flush out waste products from inside the cells. Over time, this will lead to changes in the culture medium and make it unsuitable for use. Therefore, every once in a while, usually two or three days, we need to change the culture medium. Otherwise, just wait for the cells to give you a hard time.

By so far, there must be many questions. If everything is in order and the cells are in good condition, can they just grow indefinitely in the bottle, filling it up as sand? The answer is no. To keep the cells growing in an optimal state, in addition to changing the culture medium as mentioned above, it is also necessary to monitor the changes in their number on a daily basis. This requires the use of a microscope, which is the most indispensable tool in a biology laboratory. Typically, the cells can be seen adhering to the bottom of the culture flask. The newly grown cells often side by side, like two scrambled eggs fried together on a pan. When the cells occupy 80%–90% of the bottom of the flask, it is time to consider replacing the flask. This is because once the cells reach 100% occupation and are completely packed together, they will not stack up like a ladder or only stop growing, but may even die from overcrowding. In addition to the cells that will stick to the bottom, there are also some cells that are suspended in a culture medium, such as blood cells. For these cells, daily observations of cell status and numbers are required as well.

In order to distribute the full-grown cells into new culture flasks, can you just break them off with your hands like a corncob and pour the cells into new flasks? This is such a crude approach, of course not. For adherent growth cells and suspension culture cells, the scheme of operation should be varied in the early stages and consistent in the later stages. There are two other instruments involved: an ultra-clean bench and a centrifuge. The former, as the name suggests, is a super-clean bench. It is as if the entire clean cell culture room is reduced to the size of a table, except that this time, people cannot enter. Only hands can reach in and operate. The ultra-clean bench adds an even more efficient air filtration system, so it can be almost 100% sterile. When we do cell-related work here, we can safely open the culture flasks containing the cells and change the cell culture solution. In the case of adherent cells, the sticky proteins on the surface of the cell membrane cling to the surface of the culture flask. Pulling them off rigidly can damage the cells, so it is necessary to use enzymes that deactivate these sticky proteins. Soon after the enzyme is added, usually within four or five minutes, the adherent cells will detach from the bottom of the flask and float upward as if in suspension. At this point, they can be pipetted into a tube for centrifugation and spun at high speed in a centrifuge at four or five times the normal force of gravity, causing them all to drop to the bottom of the tube. At this point, the spent culture can be discarded, and fresh culture can be added to resuspend the cells and distribute them into new culture flasks where they can continue to grow happily. We call this process cell passaging, and the rate of passaging is usually from one flask of cells to two or three. Passaging suspended cells is the same process as described above, except that it omits the step of adding an enzyme treatment, which makes it relatively simple.

After these operations, we can take off our coats, hats, gloves, and shoe covers and enjoy ourselves. Leave the cells to quietly grow in peace in the cell culture room. Remember, of course, to check on them every day. There are always unexpected events in life. Sometimes, even with the greatest care, one or two bacteria can fall into the cells and contaminate a bottle of cells. We need to get rid of it in time; otherwise, the contamination will spread, and one bottle will infect two, three, or four bottles, like the fungal nail infection.

Obviously, we have a well-established cell culture system now. However, the establishment of the system did not happen overnight. It took more than a century of precipitation, including the combined efforts of embryologists and microbiologists, to mature from scratch, bit by bit, cumulatively.

At the beginning of the twentieth century, Americans Leo Loeb and Ross Granville Harrison each began to experiment with cell culture. Loeb's first attempt was to place skin tissue from guinea pig embryos on agar containing coagulated serum. Meanwhile, Harrison used a common method of microbial cultivation, the suspension-drop method, to culture tissue of embryonic origin in a lymph suspension. Harrison then used a similar method to culture nerve fiber tissue, which was ultimately unsuccessful due to bacterial contamination. However, this allowed him to achieve the first transient cell cultures under in vitro conditions that could be observed in real-time. To avoid contamination and prolong the duration of the cell culture, he steams all the bottles and jars used for the cell culture in boiling water and autoclaves the clothes he wears. A strict aseptic technique allows cells grown in vitro to survive for weeks.

From the work of these two founding fathers of cell culture, we can see that although we now refer to these culture methods as cell culture, the conditions of the early days prevented people from getting their hands on individual cells. They were all cultured for smaller blocks of tissue. So, strictly speaking, the culture at that time should have been called tissue culture, and the use of that word has been retained in English. Nowadays, when we refer to the culture of animal cells, we can say either tissue culture or cell culture, though the latter is more common. However, we refer to the culture of plant cells, we use the term tissue culture more often. The first people to use the term "tissue culture" were Montrose Burrows of Yale University and Alexis Carrel of the Rockefeller Institute in the US. The term was introduced in 1911. Carrel, in particular, made a systematic contribution to the continuous improvement and refinement of cell culture techniques based on suspension culture. He pioneered the use of plasma as a nutrient for cell culture, then switched to serum as a substitute. He added heparin to the culture composition to prevent serum clumping. In modern cell culture, fetal cow serum has become one of the most important and widely used nutrients in the culture medium. The D flask, invented by Carrel, was the prototype of the modern cell culture flask. Based on these improvements, which greatly increased the success and efficiency of cell culture, he successfully cultured cells from various animals, including cats and dogs, and from various organs and tissues, both embryonic and adult, including skin, thymus, kidney, bone marrow, and connective tissue. In 1912, Carrel isolated a cell from the heart tissue of a chicken embryo and it was continuously and without interruption for 34 years, until two years after his death. Some have called it the first cell line, but it was controversial. Nevertheless, in honor of the first seemingly immortal cell line, WorldCom used to call Carrel every New Year's Day to notify him to see the cells.

Alexis Carrel

As you can see, Carrel's contribution to cell culture was extremely significant and he was awarded the Nobel Prize in Physiology or Medicine in 1912.

However, the reason for the award was something else. What was the reason? On June 28, 1873, Carrel was born into a merchant family in Lyon, France, sharing the same name as his father, who died when he was very young. He was raised by his mother. Because of his family's good fortune, he was able to pursue higher education and double degree programs at university. He earned a bachelor of arts at the age of 16, a bachelor of science at the age of 17, and a PhD at the age of 27. After graduating, Carrel worked in hospitals and taught anatomy courses at universities. Before the age of 30, he came to the United States to work at the University of Chicago and the Rockefeller Institute. At these two institutions, in addition to his extensive research in the field of cell culture mentioned above, he had another area of research: that of organ transplantation. In the latter field, he invented the first vascular anastomosis, which directly contributed to the development of vascular transplantation. He also discovered that cryopreservation of blood vessels can buy more precious time for surgery, benefiting more patients and opening up a new era in organ transplantation. For this, he was awarded the Nobel Prize at the age of 39. Shortly afterward, at the outbreak of the First World War, he returned to his native country to serve in the French army hospital. There, he invented the Carrel-Dakin method of wound treatment, which is still used today.

As the diversity of cell cultures in vitro increases, so do the demands on culture media. Since the early to mid-twentieth century, various researchers have extensively explored and combined various nutrients such as salts, amino acids, vitamins, hormones, and sugars in culture media according to the characteristics of the cells they were studying. They have continuously invented a variety of types of culture media. Some of them have been used as the most basic culture medium for most cell cultures, such as Dulbecco's Modified Eagle Medium. Sometimes we also refer to culture fluids as culture media, all of which refer to the nutrient solution that sustains cell growth. At the same time, substances that have nothing to do with cell growth but can prevent contamination are added to the culture fluid as well. These are antibiotics, including penicillin, which inhibits bacterial contamination, and amphotericin and mycotoxin, which inhibit fungal contamination. In the early days, very detailed experiments were carried out to see whether these antibiotics would affect the growth of the cells themselves. However, as the experiments progressed, it became clear that adding antibiotics could only prevent contamination and that it was very difficult to remove contamination that was already present in the cell culture. These studies further extended the time and efficiency of cell culture and have become an essential part of modern cell culture.

With the culture solution, cell culture can be carried out. However, there is still one problem that has not been solved, and that is the temperature. Under in vivo conditions, cells naturally exist in environments with temperatures above 30°C. For example, the human body temperature is 37°C, while the air temperature varies with the seasons and is difficult to control stably. Initially, a heated culture medium was used to simulate in vivo conditions. However, this did not work as well as it should have. At this point, it was back to the microbiologist. The second half of the nineteenth century was the golden age of microbiology, and it was during this period that modern microbiology was established. Thermostatic incubators and variable-temperature incubators were developed to meet researchers' needs for growing different microorganisms, particularly for temperature control. The first people to think of applying these to cell culture were the aforementioned Carrel. The idea of regulating the oxygen concentration in the incubator by introducing CO_2 is attributed to the New Brunswick Scientific Company in the 1960s. Interestingly, the invention of the CO_2 incubator dates back to the nineteenth century, which, due to the limited technical means of the time, was produced by lighting a candle in a tank and then placing the tank in an oven.

In the field of cell culture, a seemingly simple but very important even revolutionary step was the use of trypsin. It was first used by Peyton Rous and Jones at the Rockefeller Institute. The duo reasoned that while blood cells and sperm can be grown naturally singularly at a time in the body, cells from other tissues can only be grown together in a sticky mess and cannot be separated into individual cells. This makes it impossible to identify the characteristics of different cell types. So they thought it would be very useful for later analysis if these cells could be made into individual cells, like blood cells. At that time, plasma was already being added to cell cultures, and the fibrin in the plasma was the main factor causing the cells to stick to each other and to the culture flasks. The originator of the concept of the enzyme, Wilhelm Kühne from Germany, discovered and isolated a protein in animal pancreatic tissue in 1876 that targeted the fibrous proteins for cleavage. Hence, the name trypsin was given to it.

Peyton Rous

Based on this pre-existing theoretical knowledge, in 1916 the two men bought a ready-made dry powder of trypsin from the market. They removed the unfavorable impurities from it and sterilized it using methods such as dissolution and filtration, as described in the literature. This resulted in a yellow suspension whose activity could be maintained for two months. It was tested that when the culture fluid was removed from the cells, the addition of trypsin at a concentration of 3% quickly broke up the cell clumps while any other higher concentration would kill the cells. Cells stuck to the bottom of culture flasks quickly rounded and floated in the presence of trypsin, becoming a single floating cell similar to those found in blood cell cultures. Using a sterile centrifuge tube, they collected the digested liquid containing the suspended cells. They removed the digested liquid by centrifugation and replaced it with a plasma-containing culture medium. The suspended cells were able to grow happily on the Petri dish again. In addition, repeated suspension treatment and reapplication did not affect the growth of the cells. To see which cells could be manipulated in a similar way, they tried chicken eye choroid cells and three-day-old rat heart and stomach muscle cells. All of these gave excellent results and have been test-ed again and again. It was these studies that made cell passage possible.

Cell-passaging technology was used to try to duplicate Carrel's chicken heart cell culture experiments, which proved to be virtually unrepeatable. As a result, it has long been questioned whether his cell line can be considered im-mortal. Leonard Hayflick is among the skeptics. Born on May 20, 1928, in Phil-adelphia, Pennsylvania, the US, he was given a chemistry set as a toy when he was a boy. His father, who made prosthetic legs for a living, encouraged him to play with them so much that he set up a small chemistry and biology laboratory in the basement of their home. The first time he attended a chemistry lecture at high school, the teacher thanked him for pointing out an error in the lesson. From then on, he loved chemistry even more. Although at the age of 18, he was admitted to the University of Pennsylvania, which was on his doorstep, he was just in time for the draft and had to spend two years in the army. Fortunately, he was able to return to college and continue his studies after receiving a gov-ernment scholarship to help the Second World War veterans get back on their feet. Three years later, he graduated with a bachelor's degree in microbiology. After graduating, he found postgraduate studies too difficult and took a job as a technician in various laboratories working on bacteria. Within two years, with the encouragement of a friend, he went back to school and obtained a master of science degree and a PhD in microbiology. After completing his PhD, he went to the University of Texas in the south to learn how to culture cells and worked as a full-time cell culture technician for a few years. At the time, the cell culture world was dominated by Carrel's theory of cellular immortality. If the cells weren't grown properly or died, it was usually assumed that the technician was to blame. And from that point on, Hayflick became obsessed with cell culture.

Fortunately, a new director at the Vesta Institute in Philadelphia prom-ised Hayflick a position in cell culture. So, at the age of 30, he returned to his hometown and decided to study cells in addition to his old job of studying microorganisms. Why did he make this choice? Because it was theorized that tumor cells arise from viral infections, he wanted to extract the viruses from the tumor cells and then use them to infect normal cells. To do this, he spent two to three years cultivating human embryonic lung cells and, in total, cultivated 25 different types of lung cells. Unfortunately, after about 50 generations, the cells stopped growing in vitro, and although they did not die, they basically lay there half-dead. He then discovered that the life of normal cells is limited. They age like human beings and will not grow and pass on infinitely. Meanwhile, the cells obtained from embry will pass on more generations than those from issues of adults. In 1974, Frank MacFarlane Burnet, the winner of the 1960 Nobel Prize for Physiology or Medicine, officially named his discovery the Hayflick limit.

Hayflick Limit

Leonard Hayflick

WI-38

In 1961, Hayflick proposed his theory of the limit of cellular life. As we have just seen, normal cells live to the end of their lives in the absence of accidents. However, when an accident occurs, some cells are able to transcend this limit and literally become immortal cells. While cultivating the cells, Hayflick accidentally obtained a strain of lung cells that could be passed on indefinitely and named it WI-38. Of course, he was not the first person in history to obtain an immortal cell. Apart from Carrel's unreliable chicken cardiomyocytes, the first truly immortal cells were obtained by Katherine Sanford, Wilton Earle, Gwendolyn Likely, and others on August 16, 1948. They extracted and cultured L929 cells from the subcutaneous fat and interstitial tissue of 100-day-old mice. These cells, which could be grown indefinitely, were like a human family tree spanning generations and thus had a common name: a cell line. However, it was not the acquisition of the WI-38 cell line that benefited Hayflick, but rather the issues of ownership, attribution, and rights of use that plagued him for more than a decade with the National Institutes of Health in the midst of endless disputes and lawsuits. Of course, the dust eventually settled in Hayflick's favor.

Katherine Sanford   Wilton Earle

L929

The establishment of cell lines paved the way for large-scale cell culture, which has been widely used in the production of vaccines and antibodies. Modern biotechnology companies have also emerged on the basis of this technology. The representative cell lines used industrially are Vero cells and CHO cells. The former is called African green monkey kidney cells, which were first isolated and cultured from the kidneys of African green monkeys by Yasumura and Kawakita at Chiba University in Japan in 1962. Vero is an acronym for the Esperanto word for green kidney, which also means truth.

In the recent development and production of a COVID-19 vaccine, an independently developed and widely used type in China is based on an inactivated virus obtained by expansion of this cell line. The Vero cells can be said to be of great importance.

VERO

CHO

Theodore Puck

CHO cells, whose full name is Chinese hamster ovary cells, were first introduced in 1957 by Theodore Puck at the University of Colorado, the US. He isolated cell lines from the Chinese hamster and established them in ovaries. Because these cells are durable, they do not die easily, but grow rapidly and can be cultured in both adherence and suspension culture. They have been favored by the industry since their inception, and dozens of different types of cell lines have been derived from the first generation of CHO cell lines in an improved manner. Today, approximately 70% of recombinant proteins used in clinical disease are derived from this family of cell lines. The first modern biotech company based on the CHO cell line was Genentech. The company used this cell line to produce the first clinically applicable recombinant protein product, alteplase, for the treatment of acute myocardial infarction in 1987. This became a milestone in the production of protein drugs using cell lines. You may be wondering why this cell line with "Chinese" in its name was produced in the United States rather than China. Here's a special story from a special time in history. In 1919, Professor Hu Zhengxiang of Peking Union Medical College was researching pneumococcus. Due to limited conditions, there were no mice available, so he had to use hamsters instead. These hamsters scurried all over northern China. Professor Hu found that the hamsters were a very good research tool, just as effective as mice. From then on, the Chinese hamster became a powerful tool for epidemiological research in China. In 1948, on the eve of the liberation of Nanjing, Dr. Robert Watson of the International Medical Department of the Rockefeller Foundation in the United States escaped the war zone with 20 Chinese hamsters, traveled to Shanghai, and boarded a plane back to the United States with the hamsters. The Chinese hamsters were then successfully bred in the United States and became one of the early model animals of science.

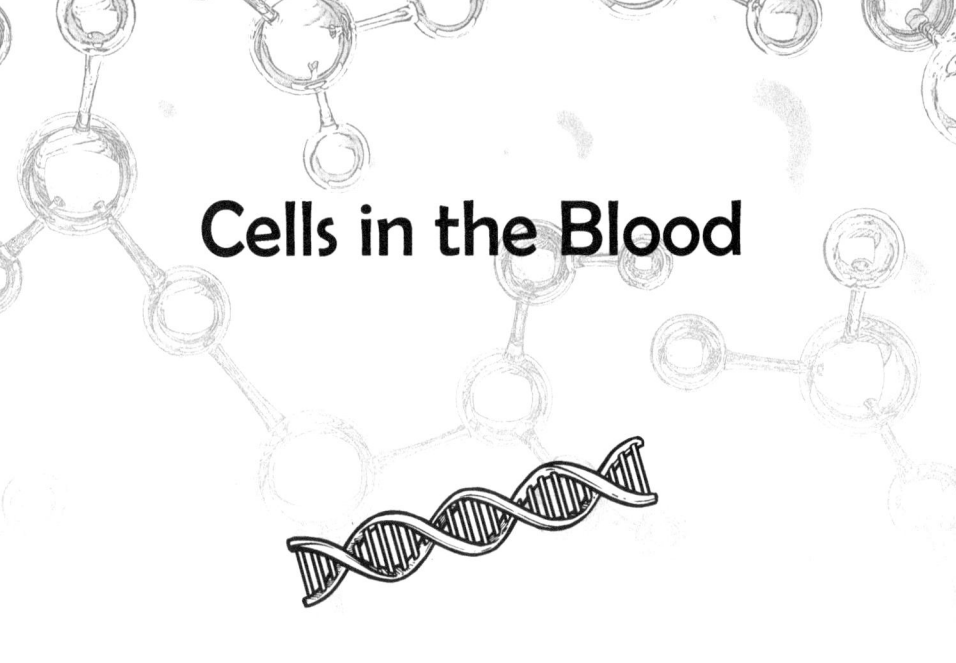

Cells in the Blood

Blood Doubt is a Japanese romantic drama that was a hit in China in the 1980s, whose heroine suffered from leukemia. The twisted plot with melodramatic storytelling techniques brought many tears to people's eyes at that time. It may have been the first time many people learned about leukemia. As a malignant tumor of the blood system, it is fatal and almost incurable under the medical conditions of the time. As a type of cancer, leukemia is extremely dangerous. Those who suffer from it often experience severe pain, bleeding, fever, and so on. Leukemia cells in the blood travel through blood vessels and throughout the body, affecting organs all over the body. While there are benign and malignant lesions in solid tissues, leukemia is only malignant and is therefore commonly known as "blood cancer."

Since leukemia is a malignancy of the blood system and blood is clearly red, why is it called leukemia, meaning white blood in Greek, not red blood disease? This is because it is indeed white, not red. To understand this, we first need to talk about the cellular composition of blood. The bright red color of blood is mainly due to the red blood cells. There are also many other types

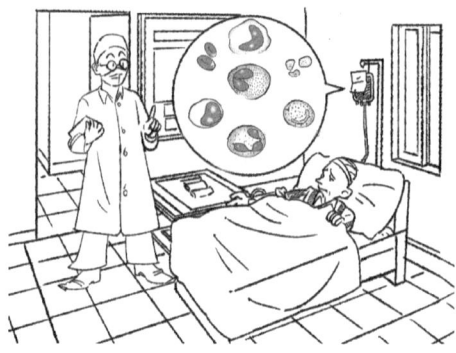

of cells in the blood, including the well-known immune cells. These cells are made up of a large number of members, such as T-cells, B-cells, NK-cells, eosinophils, basophils, neutrophils, macrophages, and monocytes. If you put this group of cells together, the color you see is white. If you really can't imagine what that looks like, just think about when you have a cut on your skin that produces pus. When you poke the white pus, a bunch of nasty white mucus comes out, and that's actually leukocyte. In leukemia, it is often the leukocytes that become sick, rather than the red blood cells. This is why the condition is called leukemia.

The discovery of leukemia dates back two centuries to 1811, when Peter Cullen, a surgeon born in Glasgow, Scotland's largest city, saw a young male patient with stomach pains and a persistent fever. At first, he used bloodletting, but after five treatments, the patient still showed no improvement. There was slight progress only after he adapted using $HgCl_2$ as a method. Cullen first observed a milky serum in the patient's blood, and although he thought it might be mainly fat, his colleagues found that the main component of this milky serum was lymphocytes. Later, in 1825, Parisian surgeon Alfred Armand Louis Marie Velpeau found an enlarged liver and spleen, as well as pus-filled blood at the autopsy of a patient with the same symptoms who died at the age of 63. In the same year, Parisian Alfred Duplay observed the same characteristics in the autopsy of a female patient who died at the age of 27. In 1829, Jacques Charles Collineau, another Parisian, used the same bloodletting procedure on a 39-year-old patient and observed the same pus in the blood. However, the patient did not have symptoms resembling hepatosplenomegaly. In 1845, the first microscopic observation of pus in a similar patient was made by the French physician Alfred François Donné. He analyzed the possibility that the differentiation of the cells had been blocked and found that there were significantly more white cells in the blood. It was also in this year that John Hughes Bennett of Edinburgh gave the name leukocytosis to the disease with these symptoms, while Rudolf Ludwig Karl Virchow named these diseases leukemia, which has been widely accepted and followed to the present day. Reviewing the entire history of leukemia detection and diagnosis, Donné and Virchow played a key role in the discovery and diagnosis of leukemia. From simple observations of clinical phenomena, such as milky pus, to the definitive identification of cell types, they contributed significantly to the field.

Alfred François Donné

Rudolf Ludwig Karl Virchow

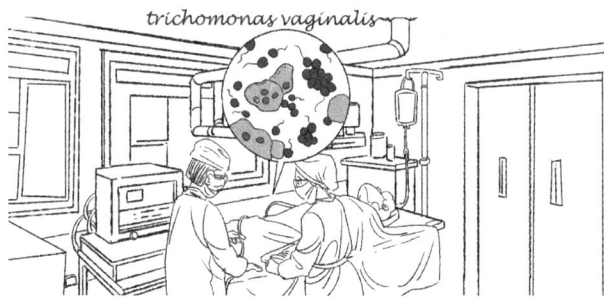

trichomonas vaginalis

Born in Noyon, France, in September 1801, Donné followed his family to Paris at the age of twenty. He studied law at the University of Paris and then moved on to medical school, graduating at the age of 28. Shortly afterward, through marriage, he joined a medical family and obtained his doctorate at the age of 30. Although his research during his doctorate focused on the use of microscopy in medicine, and he did a lot of pioneering work, the French medical community at the time was too conservative to accept new ideas. Although Leeuwenhoek and Hooke had been using microscopes to discover cells and microorganisms for almost two hundred years, Donné's work was not universally recognized and appreciated. Together with one of his outstanding students, he invented the first photoelectric microscope. This microscope used electric light instead of traditional natural light, enabling it to have a stable light source. For the first time, the projector was combined with the microscope, allowing him to better demonstrate the results of micrographic observations in real-time to others. He also used the microscope to make in-depth and detailed observations and studies of clinical specimens and made several breakthroughs, including the first observation of Trichomonas vaginalis in 1836, the first observation of blood platelets in 1842, which he mistakenly identified as fat droplets at the time, and the discovery of leukemia cells mentioned above. In addition to these contributions, he used the microscope to make detailed observations of breast milk and dairy products, leading to the establishment of the science of infant feeding and nutrition.

Virchow, for his part, was born in where is now Poland on October 13, 1821. He attended the University of Berlin in Germany and received his medical degree at the age of 22. His major scientific contribution throughout his life was attributing the occurrence of diseases to changes in cells. He lectured on the concept and tried to publish papers on it, but after being rejected by several journals, in a fit of anger he and a friend founded a new academic journal to publish his findings. In 1858, his classic work *Cellular Pathology* was published, opening up a new field of medicine. In his research into diseases of the blood system, in addition to naming leukemia, he developed the concept of blood clots. In his research into the pathways of tumor cell metastasis in the lymphatic system, the most famous concept of all is his namesake, Virchow's node, located in the left supraclavicular fossa, which has proved to be the hallmark of metastasis for a variety of tumors. In addition, he was actively involved in social and political activities. He believed that the rudimentary and unhygienic conditions in which poor people lived were mainly due to the absence of government functions. Therefore, he actively promoted the development of public health as a discipline. Interestingly, being an energetic man, another of his interests was the study of palaeoanthropology. He discovered a site of primitive human habitation in northern Germany and subsequently carried out a series of excavations, leading to the formation of a relevant association in Germany for this purpose.

There are many types of normal white blood cells in the blood, just as there are different military units in a troop with different ranks of officers and soldiers. They are fully qualified as a legion of cells in the blood. Since it is a troop, its function is to fight, whether it is a big war or a small battle. When an enemy invades the body, especially bacteria and viruses, different white blood cells have their own roles to play. If one type of white blood cell detects the enemy but cannot defeat it, it will signal to its friends of other types of white blood cells. The friends will then call to their friends, so that one calls to a hundred, and the group will rise up to destroy the enemy. The way they destroy the enemy is also different, but can be divided into two main ways: one is to eat it, digest it, and excrete it as waste, and the other is to secrete poisonous substances to poison the enemy to death. Both are extremely powerful and effective ways of fighting off almost all the invasions of our daily lives. But there are times when the enemy is too strong and the white blood cells are blinded or outnumbered. In such cases, our body needs to call in outside help, such as injections or medicines. There are also times when our white blood cells remain excited after destroying the enemy, and they just can't stop. They end up spinning out of control and throwing their neighbors around in the process. That's a problem. It will lead to the occurrence of what we often call autoimmune disease, which falls into the category of beating up on your own people. It's a head-scratcher.

The blood, as the only non-physical organ that is distributed throughout the body, contains many blood cells and can reach almost every part of the body, even the nooks and crannies. The advantage of this is that when it acts as a scout, it cannot get stuck in a dead end. This is where the white blood cell comes in, not only fulfilling the role of the body's sentinel but also fulfilling the duty of a soldier on the front line, defending our other solid tissues, organs, and the cells within them without complaint.

There are so many cells in the bloodstream. Do they all have the same function? Of course not. The body is designed to be very efficient, with very few redundant organs or tissues. Each type of cell has its own unique mission, which is, strictly speaking, just like every turnip has its own pit in the field. As we all know, one of the main functions of blood is to carry oxygen. When blood is pumped through the lungs by the heart, the tightly packed capillaries throughout the lung tissue form a vast network that allows oxygen inhaled into the lungs to burrow into the flowing blood and attach itself to red blood cells. The redness of the red blood cells reflects the fact that they contain hemoglobin, which can bind and transport oxygen molecules at a more microscopic level. Hemoglobin is more stable and does its job faithfully in the presence of iron, ensuring that the oxygen we breathe can be transported throughout the body to the cells in various tissues. Oxygen deprivation is, therefore, very frightening, especially for the brain and heart. If prolonged, it can lead to cerebral and cardiac infarctions, which are often instantaneous and fatal.

The discovery of red blood cells dates back to the time when Hooke had just discovered the cell. Jan Swammerdam, a fellow citizen of Leeuwenhoek, was born on February 12, 1637, in Amsterdam, the Netherlands. Although he studied medicine as a major in school, he was obsessed with using microscopes to observe insects in nature. To achieve the ultimate observation effect, he not only made his own handmade microscope but also crafted various fine little gadgets, such as tweezers, saws, glass tubes, scissors, to dissect insects, including butterflies, bees, and mayflies, and dragonflies. In order to better preserve the structure of various tissues or organs, he also invented various methods of dissection, such as injecting wax into the body of insects, which can preserve the shape of blood vessels, and so on. Based on all this, he used the microscope to observe real microscopic things that were neglected by other people at that time. In 1658, he became the first person to observe the red blood cell and depict its shape. He also used his observations to correct several common misconceptions, such as the fact that the leader of a bee colony is the queen bee, not the king and that muscle cells only change shape, not size, when they contract. All these observations and records led to his two extraordinary books, *The Book of Nature* and *Historia Insectorun Generalis*. For the first time, they showed people the microscopic world of insects, especially the different metamorphoses that certain insects undergo at different stages of their lives. However, in contrast to his achievements, he was a typical neet who was never financially independent and had long been relying on his parents. After losing his father's financial support, he spent the rest of his life in poverty and depression.

Historia Insectorum Generalis

Jan Jacbz Swammerdam

These blood cells travel all over the body every day, making them vulnerable to exhaustion or injury. If any accident happens, it can lead to disease. Sometimes even a tiny change can have serious consequences. In the case of red blood cells, once they start to malfunction, the first result is a reduction in their ability to carry oxygen. This condition has a big name that almost everyone knows—anemia. For those who are not congenitally anemic, a large part of the cause lies in inadequate iron intake. Why does iron deficiency cause anemia? Because the key protein in red blood cells is hemoglobin, and it needs to be bound to iron if it is to function properly. You can't have one without the other. Once iron is depleted, the hemoglobin cannot function properly, resulting in anemia. Based on this principle, the treatment of this type of anemia is mainly focused on increasing iron intake by consuming more iron-rich foods, such as pig's blood and pig's liver. If you find dietary therapy to be too slow, you can simply take an iron-rich oral solution, such as ferrous gluconate. Once the iron intake is sufficient, the red blood cells will be full of energy and happy to act as porters again.

However, there is a type of congenital anemia of which the main cause is a codon change in the genetic material that codes for hemoglobin. Once the codon is changed, the red blood cells immediately change their temperament and refuse to work, causing severe and even life-threatening anemia called sickle cell anemia. Why is it called sickle cell anemia? Because the red blood cells twist from their normal round shape into the shape of a sickle, which is used by farmers to harvest crops. Iron supplements are useless for this type of anemia because as the old Chinese saying goes, "The donkey's lips cannot be matched with the horse's mouth." They are simply not addressing the issue. So is there any hope of a cure? We'll find out in chapter 14, but first, let's talk about how the disease was discovered.

In the 1860s, the American Civil War, which ended with the emancipation of African Americans, brought with it an influx of patients in desperate need of medical care as well as an opportunity for African Americans to enter the medical profession. As the hospitals dominated by White people in the American South refused to provide good medical services to Black people, it led to an influx of poor rural Black population from the South to the North. Against this background, some northern medical schools began to enroll African American students and provide them with much-needed basic medical knowledge, thus training a number of dentists, pharmacists, and so on. Among these northern cities, Chicago was one of the most important and became known far and wide as a center of medical care and education. The headquarters of the American

Medical Association is located on the medical campus on the west side of the city, where James Bryan Herrick had once lived and worked. At the time, he was a somewhat famous attending physician at Cook County Hospital and a professor at Rush Medical College. He published several clinical case reports and wrote books, including a clinical diagnosis manual. Although his specialty was the diagnosis and treatment of cardiovascular disease, he also had an interest in diseases of the blood system. Through constant self-education, he recognized the importance of routine observation and microscopic examination of blood in clinical diagnosis.

James Bryan Herrick   *Ernest Edward Irons*

Because of its many nationally recognized doctors and researchers, Chicago has attracted not only African Americans to study but also Black people from abroad who come here in the hope of returning to their home country with medical knowledge to help their people. Grenada, a small island at the southern tip of the Caribbean, was a British colony at that time. For trade reasons, English was the official language of the island, while many people of

French and African descent were living on the island forming a population of less than 70,000, most of whom lived in rural areas and grew tropical crops. A Black man named Walter Clement Noel was born in such a place. Fortunately, his parents inherited the family's landed property and were well off to allow him to receive a full education. After graduating from university, influenced by his Black compatriots who had studied abroad in the US, Noel determined to become a dentist and set off to the US to learn dental medicine. In 1904, at the age of 20, with $70 in his pocket, he arrived in the US by ship. First, he went to New York and then to Chicago. Shortly after arriving in Chicago, however, he became very ill and had to go to the hospital to seek help for his ankle pain and breathing difficulties. The doctor who saw him was Dr. Ernest Edward Irons, an internist who had been trained by none other than Herrick. Irons performed a routine physical examination and blood and urine tests on Noel. Through these examinations, he found many pear-shaped and elongated cells in Noel's blood cell smears. In the examination report, he hand-drew the morphology of these strange cells. After receiving these reports, Herrick asked Irons to repeat the tests on Noel's blood several times at different times to verify these strange phenomena. After reviewing much of the literature, Herrick concluded that this was a completely new type of disease and named the deformed red blood cells "sickle cells." In subsequent observations, they would have liked to find out why this disease arose. However, Noel had returned home after a short period of study, and it was difficult to follow him and obtain more samples. So, in 1910, they only briefly reported the case. Since then, other researchers have gradually observed similar patients, and it has also been established that the disease runs in families. With the development of technology, sickle cell anemia became the first diagnosed molecular genetic disease, giving its discovery a place in medical history.

The sibling of the red blood cell, the platelet, is the smallest cell in the blood. Although it is called a cell, it is not a cell in the true sense of the word because it does not have a nucleus. However, it does have a complete and independent structure, so you can reluctantly call it a cell. Although platelets are small and cannot carry oxygen like red blood cells or defend the body against invasion like white blood cells, their ability should not be underestimated. When a blood vessel in our body is torn, the wound can coagulate quickly and stop bleeding, all thanks to our tiny platelets. The condition that results from a deficiency of clotting factors, causing abnormal platelet clotting, is called hemophilia. For these patients, even a small nosebleed can be life-threatening. In China, hemophilia patients are concentrated in the southern part of the country, and many cases are hereditary. The main cause of hemophilia is an abnormality in the factors involved in blood clotting after bleeding. This abnormality is often due to a mutation in their genetic material. Currently, hemophilia can only be treated with maintenance injections of exogenous clotting factors, which are expensive and ineffective, leaving many patients' families impoverished. Is there any hope for a cure for hemophilia, as there was for the diseases of platelets' sibling, red blood cells? The answer will also be revealed in later chapters.

Because blood plays such an important role and the different cells in it have so many irreplaceable functions, we need to replenish them as soon as they are deficient. For this reason, the technique of blood transfusion has gradually been mastered. In the case of excessive blood loss, the patient is unable to quickly produce enough blood cells to restore normal function. Therefore, we have to take blood from another person's body and transfuse it into the patient. Since it is impossible to find a blood donor immediately every time a blood transfusion is needed, blood banks have been set up nationally and even internationally to solve this problem. They specialize in collecting and storing blood from donors during normal times for use in emergencies. However, there is a general shortage of blood in many countries. As in a classic song sung by the Chinese singer Mao Amin, the lyrics call for the world to be a better place if everyone gives a little love. By giving a little blood, we can contribute to society. Of course, blood donation, transfusion, and storage are not so simple. There are many technical, ethical, and regulatory issues involved, and it has taken centuries of research and human sacrifice to establish the mature technology we have today.

The importance of blood was understood early on, reflected not only in Greek mythological stories but also in early Greek speculative medical theories. In Greek mythology, the iron giant Tharos, who guarded the island of Crete, had life because of the fluid that flowed through his body. At the same time, the fluid in his body became his only fatal weakness. In the fourth century BC, Hippocrates, the ancient Greek physician known as the father of Western medicine, put forward the hypothesis of bodily fluids to counter the fallacy of God-given diseases. It was believed that the human body consisted mainly of four types of fluids, including blood, phlegm, yellow bile, and black bile. The different proportions of these fluids determined the character of the human being, while an imbalance in the proportions determined diseases.

The study that really analyzes the importance of blood from a modern medical point of view begins with William Harvey. He was born in England on April 1, 1578. As the eldest of ten children in his family, it was natural for him to lead by example. He worked his way up through the ranks, winning several scholarships and earning a bachelor of arts degree at the age of 19. During his PhD studies, he studied medicine and anatomy with his supervisor and realized that in order to understand the human body, it was necessary to dissect it. This was a very valid point of view given the state of knowledge at the time. After graduating, his career progressed astonishingly and he became the royal physician to two kings. It was these unique circumstances that gave Harvey the opportunity to dissect large quantities of hunted animals, especially deers while accompanying the king on his frequent hunts. This allowed him to discover that blood in the animal's body is mainly found in veins and arteries, and that venous and arterial blood is exchanged in the lungs. In addition, the flow of blood throughout the body was mainly due to the constant beating of the heart. Thus, in 1628, the modern theory of blood circulation was established. Before this time, it was generally believed that blood circulation depended on the liver. While Harvey's contribution is now largely credited with this discovery and the establishment of the theory, there was, in fact, another important discovery that the reclusive man made in his later years that is less glamorous by comparison. He discovered that both humans and animals begin their development after the fertilization of the sperm and egg, gradually producing different parts from nothing. This is rather than what had previously been thought—the body form had already existed in the egg but had simply grown from small to large.

Of course, this little discovery was made long before Leeuwenhoek's observations and conclusions about sperm. Since there were no microscopes at that time, it was a remarkable and invaluable discovery. We can also see that the application of technology when pushed to the extreme, can often break through the limitations of the technology to make these almost impossible discoveries. In Harvey's case, it is autonomy.

Although the importance of blood was well recognized, both in the time of Hippocrates and in the time of Harvey, up until the nineteenth century, there was only one treatment for blood-based therapy, and that was bloodletting. We have learned throughout the history of leukemia's discovery that the preferred treatment for leukemia at different times and by different doctors was all bloodletting. Although bloodletting was eventually deemed useless, this did not stop people from being fanatical about it. George Washington, the first president of the United States, was bled by his doctors simply because he had a sore throat, which led to his death. In the nineteenth century, in particular, leeches were used to suck blood directly from veins to compensate for the lack of bloodletting, which therapy really flourished for a while. Bloodletting is also practiced in traditional Chinese medicine, as documented in the *Yellow Emperor's Classic of Internal Medicine* and other writings.

Having said that about bloodletting, let's talk about blood transfusion. We often say that failure is the mother of success, and this is most evident in the case of blood transfusion. By analyzing and studying a large number of failed blood transfusions, we have gradually come to realize that there are great differences in the blood cells of different blood groups and even between different people. This is how we came to understand the existence of blood groups. We now know that only blood cells of the same blood group can get along with each other. Otherwise, they will "fly off the handle." Among the many blood groups, there is one that is extremely valuable because it is rare in the population. Only one in a million people have it, so it is called "panda blood" in Chinese.

According to Harvey's records, in 1666, in a season of pleasant weather and sunshine, a group of several eminent figures gathered together, including Boyle, Thomas Willis, Christopher Wren, Hooke, and so on, all of whom were members of the Oxford Club of Experimental Physiology. At Merton College, University of Oxford, Wren demonstrated how fluids could be injected into an animal's vein and distributed throughout its body. On the evening of November 14 at Gresham College, Richard Lower demonstrated the first animal blood transfusion experiment. He cut a vein in a dog and bled it out. Then, as the dog was dying, he injected the blood of another dog directly into the dying one. The dog then came back to life, alive and kicking. The success of the experiment aroused great interest, especially at a time of conflict between different religions. People wondered if it was possible to use blood exchange to treat those who had problems in the spiritual world. With this in mind, Lower chose a madman who was believed to be possessed by the devil and made a small hole in his head on November 23, 1667. On December 12, he was transfused with sheep's blood in an attempt to treat his illness. Meanwhile, his French counterpart Jean Denis, a few months ahead of him, transfused calf and lamb blood to various patients in July. He did this with the idea of using the blood of docile animals to treat mania for therapeutic purposes. The idea was good, but the results were predictable. As a result of the successive failures of attempts to transfuse animal blood into humans, the technology of blood transfusion was put out of business and remained dormant for more than a hundred years.

Blood-letting Therapy
University of Oxford

Richard Lower

James Blundell

At the beginning of the nineteenth century, James Blundell, an obstetrician and gynecologist at Guy's and St Thomas's Hospital in England, often encountered postpartum hemorrhages that led to maternal death. So he thought he could theoretically save the lives of his patients by giving them a blood transfusion. To this end, he first carried out a series of transfusion experiments with dogs and found that both venous and arterial blood transfusions were effective. He also found that dog-to-dog blood transfusions were possible, but human-to-dog blood transfusions were not. He, therefore, hypothesized that only human blood could be transfused to humans, while animal blood could not. In 1818, he reported the first successful transfusion of blood from one man to another. Subsequently, in order to improve the efficiency and success of transfusion, he invented a transfusion device with a two-way stopper and successfully transfused other people's blood into women suffering from postpartum hemorrhage. This reduced morbidity to some extent, although it was still unsuccessful in many cases.

As to why human blood transfusions sometimes fail, according to current scientific and theoretical knowledge, it is likely that differences in blood types lead to the occurrence of clotting or hemolysis. However, it took nearly a century of fumbling to understand this common sense.

In 1875, the German physiologist Leonard Landois discovered that mixing blood cells from one animal with serum from another caused clotting and hemolysis within two minutes. In 1901, Karl Landsteiner attempted a similar experiment to discover differences between human blood. By mixing red blood cells and serum from 22 human sources, he found that the serum of some people caused the red blood cells of others to clot, while others did not. The main reason for this might be immunological differences. Accordingly, he divided the blood into three groups: A, B, and C. Serum from group C caused the blood cells of groups A and B to clot, but not those of group C. The following year, two of his students expanded the sample size to 155 people. This not only confirmed Landsteiner's previous classification of three groups but also led to the discovery of a smaller fourth group, group AB. Soon afterward, Landsteiner published his study. The results were published in German in an Austrian scientific journal, but they were not immediately recognized. In 1907, researchers in Poland and Baltimore, the US, came to the same conclusions about blood grouping, but they named it differently. In 1930, Landsteiner was awarded the Nobel Prize in Physiology or Medicine for his discovery. In 1937, the International Society of Blood Transfusion held a conference in Paris, which discussed and decided to adopt ABO as the names for blood grouping. They then have been used ever since. Besides, more complex blood groups have been discovered based on this typing.

Karl Landsteiner

In 1939, Philip Levine, one of Landsteiner's first students, discovered that hemolysis occurred in an O-type woman when she was given blood from her husband, who had the same blood type. When the woman's serum was combined with the blood cells of her husband and more than 100 other people with the same blood type, more than 80 of them formed clots. The following year, Landsteiner and his fellow researchers used the blood of rhesus monkeys to observe immunological response with rabbits and guinea pigs. Repeated these experiments and obtained the same results, which led to the discovery of antibodies in the serum and the further classification of the blood groups into Rh-negative and Rh-positive based on the negativity and positivity of the antibodies. These antibodies were named after rhesus monkeys.

Landsteiner, who made great contributions to the field of blood typing, was born on June 14, 1868, in Vienna, Austria. His father, who had a doctorate in law and was a well-known journalist and newspaper publisher, died when Landsteiner was six years old. Landsteiner was brought up mainly by his mother, with whom he was very close. After her death, Landsteiner made a mask in her likeness, which he hung on the wall of his home and reminded himself of day and night until his own death. This shows that he was a very emotional person, but also a pessimistic and misanthropic character. Even at the Nobel Prize ceremony, he did not go on stage to make a speech but asked his friend to do it for him. Colleagues around him commented that he was a person favoring plain and simple clothes and a military-like demeanor. Although he majored in medicine at university, he also minored in biochemistry and, five years after graduating, studied chemistry in the labs of a number of well-known figures, including that of the Nobel Prize-winning chemist Emil Fischer. Thus, he and Otto, the discoverer of respiratory enzymes in mitochondria, are considered to learn from the same teacher. It was these experiences that laid the foundation for his future analysis of physiological problems from a chemical point of view. After working in various laboratories, at the age of 28, he returned to the General Hospital in Vienna as a research assistant at the Institute of Hygiene. His research interests had already shifted to the study of antibodies in immunology. For more than 20 years, he worked as an anatomical pathologist, disease physiologist, and autopsy technician at the University of Vienna. At the age of 51, he moved to the Netherlands and three years later emigrated to New York, the US, where he worked at the Rockefeller Institute. He worked in the laboratory until two days before his death at the age of 76 when he was taken to the hospital after suffering a heart attack.

Today, there is no problem with matching blood types, but there are other problems. For example, in the absence of health screening of blood donors or during illegal blood collection, blood with the virus is often transfused to patients who do not have the virus, leading to cross-contamination. The spread of the Hepatitis B virus and HIV, for example, is the result of polluted blood transfusions, which was, pitifully, the exact reason why these diseases have once reached epidemic proportions in some parts of China. In addition, there have been a number of cases in history where animal blood has been transfused into humans for therapeutic purposes, causing catastrophic consequences. Although the lessons learned were painful, they gave rise to popular expressions such as "injecting chicken blood," which has persisted to the present day.[*]

---

[*] In the 1960s and 1970s, Chinese people believed chicken blood to be an effective medicine to boost energy and vitality. Therefore, the phrase "injecting chicken blood" becomes an old saying, meaning that someone is overexcited and abnormally enthusiastic to an irrational extent.—Trans.

The Ancestor of All Blood Cell Families

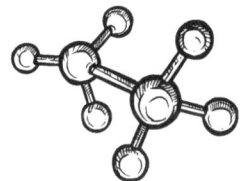

*N*ow that we've talked about the children and grandchildren of blood cells, it's time to talk about the ancestor of blood cells, the hematopoietic stem cell. All the blood cells mentioned above are descendants of hematopoietic stem cells. If the hematopoietic stem cells are seen as the great-grandmothers, then the grandmothers should be the hematopoietic precursor cells. The parents are the different types of precursor cells, and the children's generation is the immune cells and erythrocytes. For further descendants, who should be counted as grandchildren, they are cells with a very specific function or a single cell, such as plasma cells.

Once the mature blood cells have been generated, they travel through the blood vessels or lymphatic vessels all over the body and end up in different organs. So, where do the hematopoietic stem cells, which are the oldest ancestors, end up in the body? In a nutshell, hematopoietic stem cells have two homes in their life, the old one and the new one. The old home is where it is during

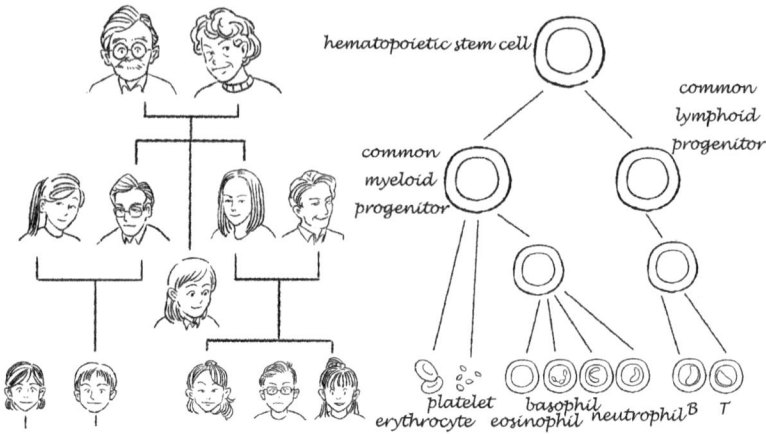

the embryonic stage, when the individual is still in the mother's body, and that place is the liver. As it grows and develops after birth, the function of the liver becomes primarily detoxification. For this, the hematopoietic stem cell has to leave its home and bring its family to the bone marrow to make a new home and end its life there. Although it is a new home, the conditions are not bad, especially the surroundings are first-class. Unlike the old house, which was surrounded only by liver cells, the new home has many more neighbors. The neighbors to the left and right have vascular endothelial cells and mesenchymal cells. In front of the house, there are not only small highways formed by capillaries but also highways formed by large blood vessels. The children of the blood cells, when they grow up, can easily follow these highways to go out and travel far away. Although a person of ambition should dare to explore the world, it is also a traditional virtue that "do not travel so far when your parents are at home." Some blood cell children, such as macrophages, in order to stay by the side of their elders, still build houses around the hematopoietic stem cells and settle down to keep them accompanied. They walk around all year round and go to visit their parents to help them from time to time. Of course, the old people will also, from time to time, advise the younger generation what to do and what not to do, so as not to go astray. This is why we say that "An elder in the family is as precious as treasure"; the joy of family here is manifested to the full.

There are so many hematopoietic stem cells collected in the bone marrow and hematopoietic stem cells can multiply to produce blood cells with different functions and abilities. So, if some cells are missing from the family due to illness and treatment, everyone will think of inviting their great-grandmother to handle the situation. However, hematopoietic stem cells, being the old ones, usually have difficulty moving around. In order to invite them to come out, it is often necessary to bring out the whole family. The back and forth of the procession is quite spectacular, ranging from millions of cells to tens of millions of cells. Such a situation is known as bone marrow transplantation. After the transplantation, the hematopoietic stem cells arrive at their new home, which has to be rearranged according to individual preferences and circumstances. Then they start to be active, one becomes two, two becomes four, four becomes eight, rebuilding their own families and villages while serving the new individuals, supporting the continuation of life while ensuring the survival of the self, and all of them are mutually beneficial and live in harmony.

Since bone marrow transplantation is so powerful, who was the first person in history to discover its function and realize this technology? It all began with the Second World War, when Japan, in a frenzy of national expansion, invaded its neighboring countries in Asia and then invaded the United States, resulting in the attack on Pearl Harbor. The attack on Pearl Harbor completely enraged the United States, which had originally remained neutral and did not take part in the war. As the world's number one power at the time, the United States had technological and economic strength that could not be underestimated. At the suggestion of Albert Einstein and others, the United States, led by Robert Oppenheimer, with the approval of President Roosevelt, launched the Manhattan Project. This project would go down in history as the creation of the atomic bomb. After several years of top-secret work, a group of geniuses came together to create the first atomic bomb in the history of humankind—"Fat Man," followed by "Little Boy," which was even more powerful than its "elder brother." These two atomic bombs were dropped on Hiroshima and Nagasaki in Japan, accompanied by huge mushroom clouds that rose instantly and destroyed the two cities, causing massive casualties. Soon after, Japan surrendered and the Second World War ended. But Japan's nightmare had just begun, and many of the Japanese who survived began to suffer from a variety of bizarre diseases that either made their lives worse or even caused them to die very quickly. Subsequent studies showed that some of these diseases were mainly because the blood cells had been exposed to nuclear radiation, which caused lesions and prevented them from performing their normal hematopoietic function, leading to low immunity or even a lack of immunity. As a result, attempts were made to treat these patients with bone marrow transplants. Politics has boundaries, but science and medicine do not. Dr. Edward Donnall Thomas of the United States was one of the scientists who worked hard to make bone marrow transplantation a reality and finally achieved partial success two or three decades after the end of the war. This ushered in a revolutionary era of stem cell therapy for which he was awarded the Nobel Prize in Physiology or Medicine in 1990. For more than half a century, bone marrow transplants have saved the lives of thousands of patients.

Edward Donnall Thomas

Joseph Murray

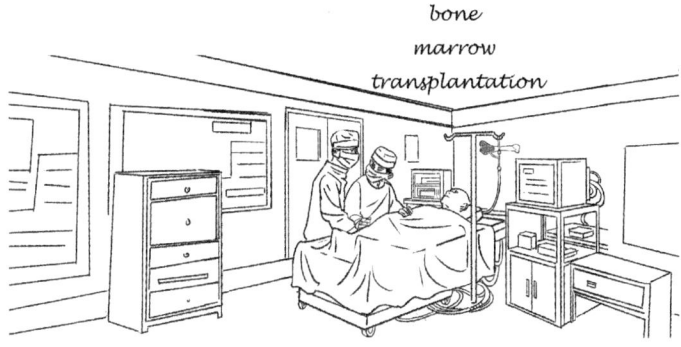

Thomas was born in Texas, the United States, on March 15, 1920. His father came here in a horse-drawn wagon 50 years ago and has put down roots ever since. Although Thomas's father had received no formal education, he did earn a medical degree. He was married twice in his life, and Thomas was born to him and his second wife at the age of 50. Having a child late in life, Thomas's father doted on him very much and accompanied him all the time as he grew up playing. Thomas was a mediocre student at school. It wasn't until he went to college that he became interested in chemistry. He pursued it hard and earned a bachelor's degree and a master's degree successively. At that time, because of the Great Depression, people didn't have any spare money to spend, although they might not worry about food and clothing. To make ends meet, Thomas worked a variety of odd part-time jobs, including serving in a women's dormitory, a job very unlikely to be available to a male student in China, where all the dormitories are guarded by aunties possessing all kinds of skills. It was through this job that he met his lifelong companion, Dottie. One snowy morning, he was accidentally hit by a snowball as he entered the women's dormitory, and their fates were sealed. In the days that followed, to support his research, Dottie, who once worked in the journalism industry, went from lab technician to lab manager, and they became a true scientific research couple. After graduating from Harvard University with a medical degree, at the age of 23, Thomas worked as a hematology intern for a year, then as an army medic for two years, and then as a postdoctoral fellow at the Massachusetts Institute of Technology for a year. After that, he worked as a resident physician and chief resident physician at the Peter Bent Brigham Hospital in Boston for two or three years. It was there that Thomas met Joseph Murray, then a surgical resident physician, who shared the Nobel Prize in Physiology or Medicine with him that year for solving the problem of rejection in liver transplantation.

Thomas's real introduction to bone marrow transplantation began with difficulties he met. At the hospital, he saw that antifolate drugs could improve the condition of leukemia patients. This made him realize that factors that activate the function of the bone marrow, such as erythropoietin, could have a good therapeutic effect. He became interested in this. However, due to the technical limitations of the time, he was unable to obtain recombinant proteins and had to abandon this research as he had no place to put his passion into action. It was at this time that studies on the protective effects of bone marrow were reported, which led him to shift his research interest and begin to recognize the clinical therapeutic value of bone marrow transplantation. First, Leon Jacobsen found that protecting the spleen enabled mice to survive lethal irradiation. Then, Egon Lorenz found that infusing bone marrow from mice was protective in irradiated mice. Invited to Columbia University's Mary Imogene Bassett Hospital in 1955, Thomas began planning human bone marrow transplantation experiments and made numerous attempts with Joseph Ferrebee using dog models. Finally, in 1957, they irradiated the whole body of a leukemia patient and then performed a bone marrow transplant using bone marrow from the patient's twin who did not have the disease. In the end, the leukemia patient went into complete remission and was even cured, which was the first report of a successful human bone marrow transplant.

However, human bone marrow transplants in the years that followed were not so smooth. Not every patient was lucky enough to have a twin sibling to donate bone marrow. Therefore, bone marrow transplants in patients without identical twins is a difficult problem that urgently needs to be solved. Two solutions have emerged from the medical community. The first is the one described by Jodi Picoult in her novel *My Sister's Keeper*: A couple's only daughter suffers from leukemia, and in order to cure her disease and save her life, on the advice and with the help of their doctors, they give birth to a younger daughter who is genetically perfect for their older daughter, an artificial twin, so to speak. From then on, the younger daughter's mission is to donate bone marrow to her sister when needed to sustain her sister's life. Of course, the ethical and social issues involved in the story are worth exploring and thinking about. However, we will not discuss them for the time being. The method described in the book is indeed compelled by the cicumstance but still a very practical approach. The second is to find out why bone marrow transplants fail so that improvements can be made. In 1963, after successfully performing the first human bone marrow transplant, Thomas began to make a name for himself. As a result, he was invited to join the University of Washington as head of oncology, where he had access to additional resources. Here, he conducted a series of studies on histocompatibility in dogs and discovered that the use of drugs could modulate the histocompatibility problems that occurred during bone marrow transplantation in animals of the same species. Soon, as other research groups in the field advanced internationally, sibling bone marrow transplantation for leukemia patients who did not have twins was introduced. The use of bone marrow from a related source that could be matched, along with the adjunctive use of drugs that regulate histocompatibility and prevent immune rejection, gradually gained success. Bone marrow transplantation has also matured over the next two to three decades and has been applied to a wide range of diseases other than leukemia.

Today, the types of diseases that can be treated with bone marrow transplantation include more than just acquired diseases. They also include those of congenital origin, affecting both adults and children. These include aplastic anemia, paroxysmal nocturnal hemoglobinuria, chronic myeloid leukemia, juvenile granulomonocytic leukemia, acute myeloid leukemia in adults or children, acute lymphoblastic leukemia in adults or children, myelodysplastic syndromes, myeloproliferative neoplasms, multiple myeloma, Hodgkin's disease, B-cell or T-cell non-Hodgkin's lymphoma, chronic lymphocytic leukemia, systemic amyloidosis, breast and germ cell tumors, renal malignancies, neuroblastoma, immunodeficiency virus infections, some autoimmune diseases, Fanconi anemia, sickle cell anemia, and dysglobinopoietic anemia. In addition to these well-established treatments, the list of diseases that can be treated with bone marrow transplantation is still evolving and will continue to grow over time.

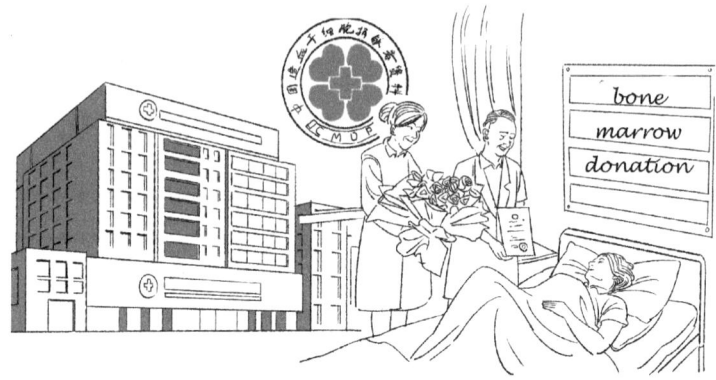

Given the high application value of bone marrow transplantation and the important role of bone marrow, bone marrow banks have been established internationally. China has also responded to international scientific development and established the Chinese Bone Marrow Bank under the guidance of the government. Similar to the establishment of a blood bank, the establishment of a bone marrow bank provides strong logistic support for occasional needs. However, compared to the number of blood donors, the number of bone marrow donors is far from the same. This is understandable because, in the early days of bone marrow donation, donors had to endure a lot of pain. After all, mechanical tools were needed to penetrate deep into the bone marrow to collect it. Collecting a sufficient amount often required a longer collection time and more puncture sites. The term "bone-marrow-deep pain" is a frightening one, and it puts many people off donating bone marrow.

As science has progressed, it has gradually been recognized that there are many types of cells in the bone marrow. When they are transplanted in a single stream from the donor to the recipient, although they all participate to a greater or lesser extent in the reconstruction of the blood system of the new individual, the most critical and important cells are still the hematopoietic stem cells. When transplanted alone, they also have a very good therapeutic effect.

Thus, after many years of unrelenting effort, researchers have finally found a way to coax hematopoietic stem cells to leave their warm nests and enter the blood vessels, just like their progeny, and begin to travel throughout the body. In this way, we can intercept them in the blood vessels of other parts of the body and no longer have to go to their home to harass and wreak havoc. This method of mobilizing hematopoietic stem cells from the bone marrow into the peripheral blood and then enriching them has now become mainstream. As a result, donating bone marrow is as easy as donating blood, and there is no longer any need to be afraid. However, many people's perceptions are still stuck in the past. The promotion of bone marrow donation should be widely publicized and the whole nation mobilized to enrich our bone marrow bank to benefit more patients.

Now that it is clear that hematopoietic reconstruction can be achieved as long as there are hematopoietic stem cells, scientists naturally thought of detecting and enriching hematopoietic stem cells in samples from different blood sources. They looked around and found that umbilical cord blood, which is often discarded in the early years, is relatively rich in hematopoietic stem cells. How much blood does a normal umbilical cord contain? About 100 milliliters. What kind of concept is that? If you put it all in a whole chicken egg, it would fill two eggs. So, how many blood cells and hematopoietic stem cells are there in that much cord blood? According to statistics, there are more than 100 million blood cells and one million hematopoietic stem cells, a ratio of about 1%. If twins are born in one fetus, the number doubles. On the basis of these figures, cord blood hematopoietic stem cells are ideal for transplantation. Of course, this can no longer be called a bone marrow transplant. Strictly speaking, it should be called a hematopoietic stem cell transplant. Because the former is so well-known, hematopoietic stem cell transplants are often referred to as bone marrow transplants, which basically means the same thing.

The first case of treatment with cord blood hematopoietic stem cells was Fanconi anemia, which occurred in October 1988. At that time, the Duke University Medical Center treated a five-year-old boy who had been diagnosed with generalized hemocytopenia at the age of two. A number of clinical signs indicated that he had Fanconi anemia, including developmental delay, slowed growth, six fingers on the left hand, an absent left kidney, and hypospadias. If left untreated, there was a risk of cancer developing, which could be life-threatening. Fortunately, his parents were not consanguineous and there was no family history of blood-related genetic disorders. In addition, his mother had become pregnant again in the previous June. Whether intentionally or unintentionally, we do not know, but in February of this year, his mother successfully gave birth to a baby girl. Initial tests showed that this little sister was a healthy baby without Fanconi anemia. Therefore, two large bags of umbilical cord blood and one bag of placental blood from her birth were frozen. Two weeks before the cell transplant for this little boy, the cord blood and placenta blood were flown from Indiana University School of Medicine to Duke University using cold-chain shipping techniques that kept the blood at $-175°C$. On the day of transplanting, the frozen blood was warmed and revived. Studies showed that 82% of the cells were still alive, and some were sent to Paris for testing. The tests also showed that they still contained hematopoietic stem cells. In order to increase the success rate of the transplant and minimize infection, the little boy was placed in a separate and special room more than a week before the transplant. This room, such as the cell culture room mentioned before, allowed for air filtration and irradiation of all surgical items and equipment. That came into this room. He was also given a series of antibiotics or injections to prevent bacterial, fungal, and herpes simplex virus infections. The day before the transplant, his chest and abdomen were given a dose of radiation, and his lungs and liver were covered to prevent damage to other organs. Two hours after the cord blood transplant, the little boy experienced acute symptoms of discomfort such as tremors, fever, and elevated blood pressure. These symptoms quickly resolved. A month later, he experienced the early side effects of rejection and liver dysfunction and gradually returned to normal. Six months later, blood cells and hematopoietic stem cells from his sister were detected in both his peripheral blood and bone marrow, proving that cord blood transplantation has the same hematopoietic stem cell transplantation effect as bone marrow transplantation.

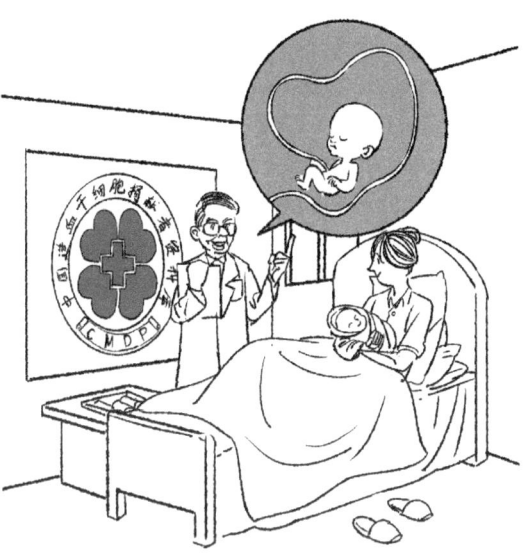

As with the establishment of blood and bone marrow banks, we have established cord blood banks in recognition of the immense value of cord blood. Human reproduction is infinite, with tens of millions of babies born each year, each with an umbilical cord, so the collection of umbilical cord blood would be infinite as well. In addition, the umbilical cord was originally discarded as a useless object, and collection is non-invasive and causes no discomfort to either the mother or the baby. The size of cord blood banks will definitely exceed that of bone marrow banks and will become an important source of hematopoietic stem cells. At present, there are two types of cord blood banks in society: the public cord blood bank managed by the government, which relies on donations from families and is in the public interest; and the private cord blood bank set up by companies, which relies mainly on family-funded storage and is equivalent to handling cell insurance. However, the probability that an individual will need cord blood stored at birth during his or her lifetime is extremely low. In addition, the transplantation of hematopoietic stem cells must be adapted to the patient's body weight. For an adult weighing more than 50 kilograms, the amount of cells needed normally equals cord blood from four or more eggs, as mentioned above. We can all do the math. Have you saved enough cord blood for yourself? If everyone can agree to donate to a cord blood bank to create universal sharing and reciprocity, the effect will definitely be even better.

Up to this point, readers who have carefully read the previous part of this book and who like to use their brains will surely ask themselves, since it was said before that cells can be grown from few to many, why not directly cultivate hematopoietic stem cells for transplantation treatment instead of establishing bone marrow and umbilical cord blood banks? In theory, for hematopoietic stem cells of different blood types, as long as a small amount of each type of cell is obtained and cultured on a large scale, an abundant and unlimited supply of hematopoietic stem cells can be produced. This would allow any patient needing a transplant treatment to access the cells anytime and anywhere. This was the original wish of scientists, but as research into hematopoietic stem cells progressed, it became clear that while the idea was good, the reality was cruel. There is a paradox in the culture of hematopoietic stem cells: if the hematopoietic stem cells want to maintain their old ancestral status, they cannot proliferate, which we call the "quiescent state." Once a cell has been cultured into the proliferative state, the quiescent state is lost. Although it still looks like a hematopoietic stem cell, it has lost its properties as an old ancestor. It is, therefore, extremely difficult to culture hematopoietic stem cells. Even so, many researchers, both domestically and internationally, remain true to their original aspirations and persevere in the face of challenges. With the determination that difficulties are challenges and breaking through bottlenecks will bring opportunities, they tirelessly devote themselves to the culture of hematopoietic stem cells. It is expected that in the near future, a pure hematopoietic stem cell bank will be able to replace cord blood banks and bone marrow banks, bringing hope to more patients.

Those of us who have studied dialectical materialism know that there are two ways of looking at everything. These mentioned above are all the positive aspects of hematopoietic stem cell transplantation, and finally, let us talk about the negative aspects. Let's start with three criminal cases in history where it was the forensic pathologists who solved the mystery instead of the police. In November 2000, to find out the criminal of a rape case, the police analyzed the semen left in the victim's vagina and the bodily fluids left in her underwear for genetic material and compared the results with the National Genetic Information Bank. The results of the comparison showed the genetic information of two men, one of whom turned out to be her own brother. What's going on here, are her own relatives really involved in this case? The other case was the identification of kinship between two bodies that had not had time to escape a fire and tragically perished. Both victims were children, a boy and a girl. The bodies were severely burned, making it impossible to identify them from their appearance. For this reason, it was necessary to test their genetic information and match it with their parents in order to verify their identity. The results of the tests were very surprising. The blood taken from the boy's heart tested clean and it was certain that he was the child of his parents, while the test results of the little girl were negative, so could it be that she was not their daughter? The third occurred in 2004, a sample of semen from a suspected murderer was found at a crime scene in Alaska, the United States. When the sample was analyzed for genetic material and compared with a genetic database of criminals, a match was indeed found. The problem is that this suspect had no time to commit the crime because it is 100% certain that he was serving a prison sentence at the time of the crime. So if he was not the killer, why was his genetic information present at the crime scene? Through detailed questioning of all the participants and suspects in the above cases, it was found that whether the victim's brother in the first case, the little girl in the second, or the suspect in prison in the third, they all had a common feature related to bone marrow, that is, they had either undergone bone marrow donation or bone marrow transplantation, which resulted in the appearance of genetic information in their bodies, or in the bodies of other people, that was genetic information belonging to others. This accidental genetic information is carried by hematopoietic stem cells, which makes the whole case complicated to analyze and often leads to wrongful convictions if not followed up. It is worth noting that the reason why the donor's genetic information was also found in the semen was not that the hematopoietic stem cells had turned into sperm but because of the presence of a large number of immune cells in the semen, which originated from the hematopoietic stem cells.

It is not yet clear whether hematopoietic stem cells can turn into sperm so that the bone marrow donor can influence the recipient's offspring at the genetic level. In the field of forensic testing, this strange and unusual phenomenon has come to be known as the "chimera phenomenon," which refers to an animal called chimera from Greek mythology, which had the head of a lion, the body of a sheep, and the tail of a snake. If they had been discovered in China, I think they would have been named after one of the beasts in the *Classic of Mountains and Seas*, because there are so many mythical beasts recorded in this book that are similar to chimera, but derived from Chinese mythology, such as Lingyu, Kaiming Beasts, Loongyu, Bingfeng and Longzhi, and the list goes on and on.

chimera

Stem Cells

$\mathcal{H}$aving discussed hematopoietic stem cells, you may still be wondering what kind of cells the stem cells are. According to the literal meaning of the word, laypeople may think they are cells that can do anything.* However, this is a complete misunderstanding. The meaning of stem comes entirely from the stem of the tree trunk. Just as the giant tree can only flourish because of its trunk, so too does the stem in stem cells. Many cell types are derived from stem cells. In this sense, the stem is like the mother of a mother or the ancestor of an ancestor. That's why stem cells are sometimes called "progenitor cells."

What can stem cells do? The classical biological definition is a cell with the ability to self-renew and the potential to differentiate. How can this be understood? To use an inappropriate analogy, if a person is able to give birth to a baby and can also turn oneself into two or four, then we can call such a strange

---

* Stem cells in Chinese are called "gan xibao" (干细胞). Despite the meaning as stem, the word "gan" (干) also represents the action of working. —Trans.

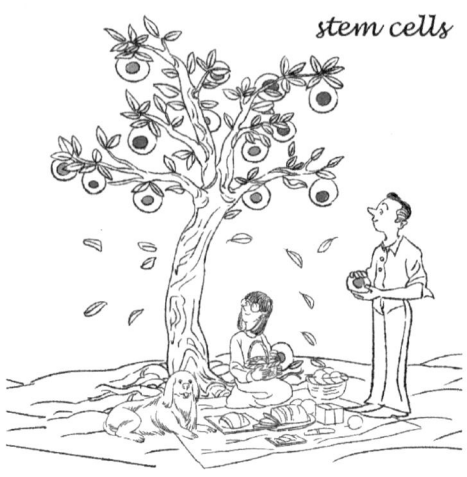

stem cells

person a "stem person." Of course, it's impossible for a human being to do that. Nor can any other animal. But a cell can, and that is a stem cell. So, in general terms, if a piece of tissue or an organ is missing a piece of flesh, in the presence of stem cells, one stem cell becomes two stem cells, and two stem cells become four, eight, sixteen, and so on. These cells can be further transformed into cells that build three-dimensional tissues and participate in repairing the damaged parts, thus achieving the purpose of regeneration. This is the attraction of stem cells and why they are so sought after.

Since stem cells have such powerful magic, is it possible to use stem cells directly to treat diseases and achieve what we want? To answer this question, we need to understand how many types of stem cells there are because different stem cells have different functions and we should not mix them up. In terms of time, stem cells include totipotent stem cells, pluripotent stem cells, and monopotent stem cells from early to late stages. In terms of space, most stem cells can be generalized as adult stem cells, which can be called different types of stem cells according to the location of different tissues or organs. For example, the stem cells of the nervous tissue are called neural stem cells, and the stem cells of the blood system are called blood stem cells. The above classification of stem cells is often intertwined. For example, monopotent stem cells are mainly adult stem cells, while totipotent and pluripotent stem cells are mainly found in the embryonic period or even earlier, at the beginning of life before it is fully established.

When it comes to stem cells, another concept that needs to be mentioned is development. Anyone who has a child at home can easily understand that development means the growth of a child from a small child to a big child, including growth in size, body mass, the appearance of secondary sexual characteristics, and so on. This is not wrong at all. There is nothing wrong with that. It is completely correct, but the development we are talking about here starts with one of the most specific cells: the fertilized egg. What is a fertilized egg? Where does it come from? When a father and mother fall in love and start a family, the father provides the sperm and the mother provides the egg. When these two bodies meet and embrace, the fertilized egg is formed. This is the origin of all development. After this, it undergoes all sorts of magical changes that produce different types of stem cells, tissues, organs, embryos, and conscious-thinking individuals.

So, at what stage of development do the totipotent and pluripotent stem cells we mentioned earlier arise? When a fertilized egg is created, one cell becomes two, two becomes four, and so on. Then there are 8, 16, 32 ... Immediately after that, the functions of these cells are divided. Some cells continue to develop into the outer layer that surrounds the other cells, while some cells hide inside these protective layers. These hiding cells are pluripotent stem cells, and they have a very good name—embryonic stem cells. We will explain why it is so famous in a later chapter. The cells that develop into both a protective outer layer and an inner embryonic stem cell are called totipotent stem cells because they can really develop into all the cells needed for a mature individual. The further development of the outer cells is not particularly fancy; they simply provide protection and nourishment. The inner embryonic stem cells, on the other hand, display all the abilities of Sun Wukong's seventy-two transformations and eventually develop into cells with different functions, including adult stem cells.

As the cells in the embryonic stage increase dramatically, accumulating and overlapping, they transform into an embryo and gradually possess a conscious mind. This is what makes a fetus. In the womb, the fetus absorbs nutrients through the placenta and grows until it leaves the mother's body. With a cry, it comes into the world as an independent living being and becomes a baby. From this point on, we define all the stem cells in her or his body as adult stem cells throughout the adolescence, youth, middle age, and old age. This is because, within these periods, all tissues and organs have basically been formed, and there are only more or less stronger or weaker differences in function. The stem cells present in each tissue or organ, i.e., adult stem cells, are also called the stem cells of that tissue or organ, as in the examples given above.

There are two ways in which adult stem cells regenerate: one is endogenous and the other is exogenous. The so-called endogenous source is the adult stem cells present in the body. When there is damage to the tissue or organ in which these stem cells are located, they begin to work. They replicate themselves to increase the number of cells and differentiate into other cells to plug the wound and replace the lost cells. At the same time, they release a multitude of chemical signals from the body, just as a male animal secretes sex hormones to attract females. These signals recruit cells that it cannot produce on its own to help itself and its descendants repair the damaged parts together. This achieves the purpose of regeneration. Exogenous stem cells mainly refer to the artificial importation of adult stem cells to increase the number of stem cells involved in repair or to fill stem cell vacancies in the body to achieve the purpose of repair or improve the speed of repair.

If adult stem cells at later stages of development have such a strong regenerative capacity, would embryonic stem cells at earlier stages of development have an even stronger capacity, and would embryonic stem cells given exogenously have a better repair rate than adult stem cells? We say that it is obviously not. Logically, you can argue that way, but in fact, it is completely impractical. Because embryonic stem cells only appear in the embryonic stage of natural development, if they are forcibly transferred to adult tissues, they won't work the way we want them to—after all, a melon forced to grow is not sweet. They will entertain themselves in the transplanted tissues and continue to exercise their superior ability to reproduce their offspring, creating their own little world.

This little world will be filled not only with cells that are needed for localized tissues, but also with other cells that shouldn't belong there in the first place. This threatens the native cells of people, and instead of helping, it will be of no benefit at all. If that is the case, shall we say embryonic stem cells have no use in regenerative medicine? We have just discussed that adult stem cells work both endogenously and exogenously. If embryonic stem cells play a negative role in the endogenous case, how do they behave in the exogenous case? According to research and theoretical considerations, we believe that embryonic stem cells can develop into different types of cells under exogenous conditions, which in biology are often called in vitro culture conditions, just as they develop in vivo. Therefore, we can use embryonic stem cells under in vitro conditions to transform them into the type of cell we want to obtain. If we have difficulty obtaining these types of cells from the body, they are used as exogenous cells that are transplanted into injured tissues or organs for regenerative repair. For this reason alone, embryonic stem cells are like a gift from God with unlimited applications. This is why they have been pursued with such enthusiasm throughout history.

However, you must think that stem cells are everywhere, that all tissues should have their unique stem cells, and that all stem cells should be beneficial, right? This statement is both true and false. Why is that? True refers to the fact that stem cells are not only present in normal tissues, but they are also present in mutated and demonic tissues such as tumors. However, the more of these stem cells there are, the worse they are, not the better. False refers to the fact that stem cells are not present in all normal tissues or organs, and one of the most amazing organs of all is the heart. The discovery of cardiac stem cells was once considered the most exciting discovery of all time. Because it has long been accepted that the heart cannot regenerate or functionally repair itself once it has been damaged by a heart attack, cardiovascular disease has always had a high mortality rate and is one of the top three causes of death in humans. If cardiac stem cells exist, it is theoretically possible to regenerate the heart after injury. It would just be the case that we have not yet found the recipe to unlock the magic. However, nothing is perfect, and the available evidence suggests once again that the existence of cardiac stem cells is a beautiful lie. In other words, there are no cardiac stem cells in the heart. Hope and despair are thus in and out of the world, playing with people.

The importance of the heart in the human body cannot be overestimated. In the case of various mechanical injuries, as long as the heart is not injured, we often think that the person can still be saved. When we watch various television dramas or films, we often hear doctors say after an operation, "It was only one centimeter from the heart; it was too close." This is obvious. In normal life, for the ordinary person, heart disease manifests itself in two main ways: one is heart failure, where the heart's ability to contract and diastole is impaired, causing blood to pool in the blood vessels; and the other is a heart attack, or myocardial infarction, which is the acute death of cardiomyocytes due to blockage of the heart's blood vessels, accompanied by angina pectoris and often associated with a very high risk of death. Faced with these two cardiac killers, stem cell therapy has therefore focused on two main directions. The first is the use of stem cells to replace deficient cardiomyocytes, and the second is the use of stem cell secretions to provide cardioprotective functions.

In 2000, some scientists reported that hematopoietic stem cells could not only produce many kinds of blood cells in the blood but also penetrate into the heart and turn into cardiomyocytes when the heart is damaged, thus achieving the purpose of repairing the heart. The problem is that this result was controversial and has not been confirmed until now. The following year, American Piero Anversa found cardiac stem cells in the heart. This means that such stem cells can directly differentiate into cardiomyocytes, which can constantly replace lost cardiomyocytes. This sensational discovery has earned him numerous awards, hundreds of research projects, and numerous research grants, making him one of the most important figures in the field of stem cells and, of course, one of the best in the field of cardiology. A new direction in cardiac stem cell research under his leadership became a hot area of research in the United States and around the world, attracting thousands of researchers to follow. Soon after the discovery, however, controversy arose as one researcher after another reported that they could not find the so-called cardiac stem cells.

It wasn't until 2018 that cardiologists from around the world published a joint paper stating that heart stem cells don't exist and that all research on them is a hoax, putting an end to the controversy. So we can't help wondering why this hoax has been going on for 17 years if cardiac stem cells don't exist. Ultimately, it is because Anversa himself, as an authority in the field, had adopted a policy of suppressing research results that differ from his own. This made it difficult to publish questionable papers. Second, many researchers still adopt an attitude of blindly trusting and worshipping authority, which naturally makes it difficult for them to detect hidden problems and tricks in the process. Here we return to the old problem of scientific research. We must dare to ask questions and challenge authority, but at the same time, we must do so with justification and not with slogans. Only in this way can we make new discoveries and promote the progress of science.

When the story of the fake cardiac stem cells broke, it quickly sent the field of research into an icy valley. However, there is reason to believe that while stem cells may not be able to help repair the heart, they are worth studying in depth for cardioprotection. Both hematopoietic stem cells and other types of stem cells, such as mesenchymal stem cells, which we will discuss later, have been shown to have beneficial effects on cardiomyocyte protection and vasculo-protection after cardiac injury. In addition, although the heart is dominated by cardiomyocytes, it also has a large number of blood vessels. The vascular endo-thelial cells that make up the blood vessels contain cardiovascular endothelial stem cells, which are still very helpful in repairing damaged blood vessels after a heart attack. Therefore, since cardiomyocyte repair is a difficult path to take, it is also beneficial to take a circuitous route to achieve an auxiliary protective function by repairing blood vessels.

Like the heart, the brain, or, in other words, the entire nervous system, is extremely important to humans and other animals. Once damaged, it is not a trivial matter, either killing or severely disabling. In addition, the main cells that make up the brain and nerves are neurons, just like cardiomyocytes in the heart. They do not have the ability to regenerate once damaged or dead. This piece of neural tissue, even if a piece of flesh is missing, is very difficult to repair and regain function. However, the nerve is much more fortunate than the heart because not only have neural stem cells been shown to exist, but they have also been used to treat certain neurological disorders. Therefore, the future is really promising.

Sally Temple

Fred Gage

Brent Reynolds

To this day, there are two schools of thought that debate whether new neurons are naturally produced to replace missing cells after the loss of neurons in the nervous system. One school of thought is that newborn neurons are produced only during the embryonic period and that no new neurons are produced after birth, especially in adulthood. The other school of thought is that new neuron production occurs all the time, whether in embryos, young children, or in adulthood but becomes less frequent with age. Whatever the debate, the discovery and successful isolation of stem cells in the nervous system suggest, to some extent, that nervous system regeneration can indeed occur. Between 1989 and 2002, Sally Temple, Brent Reynolds, Jonathan Flax, and Fred Gage successively isolated and cultured neural stem cells from embryonic mice, adult mice, human embryos, and adult brains. However, due to limited technical conditions and a lack of understanding of the characteristics of neural stem cells, researchers mainly use the suspension culture method to cultivate the cells isolated from the brain. This is because it is difficult for cells such as neurons to survive, so the cells that survive are theoretically stem cells. Neural stem cells like to clump together as they grow, and as the number of cells increases, they clump together to form an easily recognizable ball, hence the name neurosphere. To prove that neural stem cells are a type of stem cell, in addition to the ability to form a ball, it is also necessary to see whether they can behave like hematopoietic stem cells in general, with the characteristics of the identity of the ancient ancestors, such as roduce offspring and grandchildren. The progeny obtained from the differentiation of neural stem cells is not as large as the hematopoietic stem cell family, but it is not small either. In general, it includes three groups of descendants: neurons, astrocytes, and oligodendrocytes. The morphology of each progeny cell has been described in the previous sections. Among them, neurons can be divided into different types according to the different parts where they are located and the different functions they perform. Therefore, they can definitely be considered as a large and unique family!

Now that the existence of neural stem cells is known, and a large number of neural stem cells can be successfully isolated and cultured to obtain a large number of neural stem cells, how can these cells be made to serve us for the purpose of regenerating the nervous system after injury? To further explain neural stem cells for nerve regeneration, it is first necessary to explain two terms—psychiatric illness and neurological illness—to avoid confusion. The former refers to disorders that affect mood, thinking, and behavior and often do not involve the physical dimension, so they are outside the scope of our discussion. The latter refers to lesions of the nervous system that can cause various functional disorders that affect daily life and can even be life-threatening. Neurological disorders are caused by both congenital and acquired factors, with a wide variety of triggers and treatments that vary from disease to disease. Neural stem cells are not evenly distributed throughout the nervous system but are mainly confined to a few specific locations. Just as hematopoietic stem cells prefer to live in the bone marrow, neural stem cells prefer to stay in two places in the brain called the subventricular zone of the lateral ventricles and the subgranular zone of the dentate gyrus of the hippocampus. When nerves in other parts of the brain are damaged, the neural stem cells in these two places can migrate all the way from their home to the site of the injury, following the path laid down by the nerve fibers, on the one hand, it secretes trophic factors to calm the less severely damaged neurons and prevent them from losing control of their emotions in response to shock; on the other hand, they will differentiate into the lost neurons or other types of cells and play the role of replacement players to achieve tissue regeneration and functional repair of the damaged area.

For injuries that are so far away that they are difficult to reach, either by walking or running fast parts, it is basically unrealistic to try to move the neural stem cells by the walking method step by step. Therefore, the method of neural stem cell transplantation can be considered for the treatment of this type of neurological disease. The good thing is that, unlike hematopoietic stem cells, once they are isolated from the body, they can be easily cultured in vitro and they can grow as many as you want and they grow quite fast. So for the time being, there is no need to set up a neural stem cell bank. Next, we will present three typical neurological diseases, all of which can be treated with neural stem cell transplantation, and good results have been achieved in some patients.

Spinal cord injury is a relatively common neurological condition, often caused by external forces that compress and break nerve fibers in the spine, resulting in paralysis. For example, the Chinese gymnast Sang Lan and the famous film star Christopher Reeve, who played the role of Superman in the United States, were paralyzed as a result of training and horse-riding accidents and have been sadly confined to wheelchairs for the rest of their lives. With the ability to transplant neural stem cells to the site of the injury within a short period of time after the injury, along with other gels or scaffolding materials to prevent the cells from being lost with the spinal fluid at the transplant site, it would possible for both of them to get out of their wheelchairs and get back on their feet and walk again.

Parkinson's disease is a neurodegenerative disease that often affects older adults. As we age, nerve cells degenerate or become diseased, resulting in the

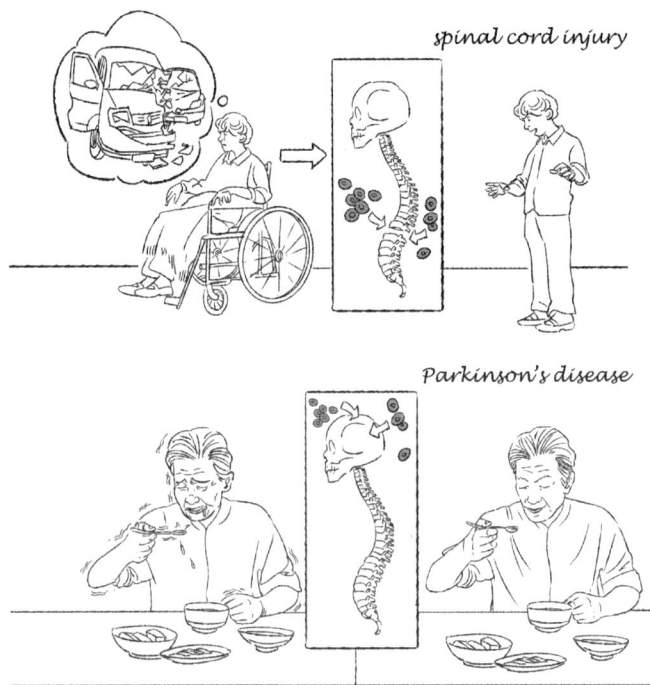

spinal cord injury

Parkinson's disease

loss or death of a type of cell called a dopaminergic neuron, which secretes less dopamine and causes older people to experience involuntary stiffness or tremors in the muscles of the hands and feet. This may be the reason why when some people eat with a spoon, their hands keep shaking, making it difficult to get food to their mouths. In order to treat this disease, the neural stem cells need to be slightly manipulated in vitro to allow them to differentiate slightly and turn directly into dopaminergic neurons or their precursors, and then these cells can be injected into the site of the lesion, which theoretically can be used to achieve the effect of treating this type of disease.

The third disease is neurological tumors, particularly brain tumors such as gliomas, which are very malignant and for which there are no good targeted therapies. Taking advantage of the fact that neural stem cells can migrate and prefer to crawl toward the site of injury, now there are scientists have neural stem cells as a "van," loaded with drugs, and then these cells are injected into the brain. The drug-loaded neural stem cells will be directed to the tumor site, and once they reach the predetermined location, they will unload the "cargo" and kill the tumor cells, thus achieving precise treatment without harming other normal cells.

Of course, although the treatments for these diseases that have been introduced so far have already undergone clinical trials on some patients and achieved more or less the expected results, the overall situation is not very stable, especially in some patients there is no therapeutic effect, so there is still a long way to go before it is finally popularized among the public. In any case, it is a promising path, and we believe that in the near future, with a deeper understanding of the disease and a more comprehensive understanding of the characteristics of neural stem cells, we will be able to realize the full victory of neural stem cell transplantation in the treatment of neurological diseases.

Having talked about hematopoietic stem cells and neural stem cells, let's talk about mesenchymal stem cells. If you go to the US government's clinical registry website, you will see that in terms of the number of clinical trials that have been done, in addition to hematopoietic stem cells, which have been done thousands of times, the second most common stem cell, mesenchymal stem cells (MSCs), have also been done thousands of times. Neural stem cells, on the other hand, have only been done a few dozen times. Currently, MSCs are the most common stem cells on the market. Clinical trials of MSCs transplantation are in full swing in hospitals large and small, public and private, and the types of diseases being treated are diverse. Quite apart from the fact that it is difficult for the layman to understand, it is like looking at flowers in a fog for those of us involved in stem cell research. So, why are MSCs so popular? Do they really have the power to cure all diseases?

In 1991, American researcher Arnold Caplan first named a group of stem cells that had been isolated from bone marrow by other researchers three years earlier as MSC. He showed that these cells had the ability to differentiate into cartilage, bone, and fat. Over the next ten years, research groups around the world used a variety of modern techniques to characterize these cells and find ways to separate them by their appearance. As studies progressed, researchers

Arnold Caplan mesenchymal stem cells

increasingly concluded that these cells were not stem cells. At the 2006 annual meeting of the International Society for Cellular Therapy, it was unanimously recommended that their name be changed to pluripotent mesenchymal stromal cells. Despite having an official name, people didn't buy it, and the name MSCs became so entrenched that even when Caplan himself suggested a name change, such as calling them medical signaling cells, it didn't help. Because there is no single standard, the sources of MSCs are endless as more people get involved in research. In addition to bone marrow, these cells have been found and isolated from adipose tissue, umbilical cords, placentas, amniotic membranes, teeth, and women's menstrual blood. In addition, the differentiation capacity of MSCs is no longer limited to the three cell types mentioned above. They can also be differentiated into cardiomyocytes, hepatocytes, neuronal cells, myocytes, and endothelial cells in blood vessels. The only thing you can say is that you can't imagine it. There is nothing it can't do. Because of these unclear reports, the function of MSCs was immediately deified. Especially when it is fueled by the capital market, it can be said that money makes the devil work. Weak functions are called very strong curative effects, and even functions that are not there can be said to be effective. In this way, after 30 years of frenetic development since their discovery, MSCs have already become a universal stem cell, with more than 30,000 research papers on the subject and applied to the treatment of a wide range of diseases, from psychiatric to neurological, from pulmonary to gastric, from intestinal to hepatic, from dermatological to immunological, and so on.

An under-studied cell, used to treat so many diseases, may offer many unexpected surprises, but it is often fraught with danger. Market self-purification alone often costs lives. From this point of view, professional researchers and those who formulate and regulate government policy in this area are bound to be ahead of the market and the general public. From current research data, it is clear that MSCs work in two ways. On the one hand, they have an immunomodulatory function, inhibiting inflammatory responses, promoting wound healing, and inhibiting cell death by secreting a variety of factors. However, whether there are differences in the functions of cells from different sources, how to standardize them and how to accurately apply them to the treatment of different types of diseases are major problems that need to be solved over a longer period of time. Otherwise, MSC treatment will always be a pot of porridge and it will be impossible to say whether it is effective or not! On the other hand, MSCs need to be more carefully categorized, with different subtypes of cells being used for different diseases. For example, recent studies have shown that there is a type of skeletal stem cell in this group of cells that can differentiate into bone and cartilage cells, but not into fat cells. Therefore, this type of cell will definitely be an important seed cell for future cell therapy after joint injuries.

The stem cell and regenerative medicine we often hear about is a very broad concept. When put into practice, we may often hear the professional name "stem cell and tissue engineering." In regenerative medicine, stem cells can only be described as seed cells. When it comes to repairing some tissues or organs, it is like a farmer planting seeds in the spring, sowing them, and harvesting them in the autumn. But for some special plants, such as grapes and cucumbers, it's not enough to just plant the seeds and leave them alone. You have to build a trellis for them to grow happily. The same is true in the field of regeneration, but the scaffolds needed for stem cell growth are not ordinary branches or bamboo strips. Instead, they rely on the development and constant renewal of modern materials science, which we call scaffold materials. These materials must have good tissue compatibility, allowing the stem cells to climb up and live in harmony with the tissue. They should not be toxic and at the same time have a certain degree of toughness and degradability. They should also have a certain amount of internal space, and the size of the gap can be adjusted according to the size of different stem cells. This way, the cells can live and work happily inside. In these materials, some of the gifts of nature are used, the most common being silk. Silk can not only be used to make clothes but can also be melted and then re-sprayed into a new shape, which serves as a first-class stem cell scaffolding. The other method uses synthetic technology to mimic the material and is then smeared with a layer of cell secretion containing various proteins and sugars, which the stem cells like a lot.

MSCs have the ability to differentiate into chondrocytes, and the combination of these cells with scaffold materials has shown great promise in orthopedic regenerative repair. In 1997, Cao Yilin, a scientist in China, isolated soft osteoblasts and cultured them in vitro. He obtained a sufficient amount of cells that were co-incubated with biodegradable materials made of human ear-shaped scaffolds for a period of time. The chondrocytes not only climbed all over the scaffold but also drilled into the scaffold inside. He then transplanted this ear-shaped scaffold containing chondrocytes into the subcutaneous skin of a hairless mouse, creating an artificial ear growing on the back of the mouse. From a distance, it looks like the villain in the Chinese cartoon *Sheriff Black Cat*, a rat with one ear. This is extremely shocking. This technology has not just stayed within the laboratory and has had a significant impact. After two decades of hard work, Cao and his team established the National Tissue Engineering Center. Together with the Plastic Surgery Department and other departments affiliated with the Shanghai Jiaotong University School of Medicine, they successfully transplanted artificial ears made from stem cells into children with defects in their outer ears. This not only significantly improved the patient's quality of life but also made a significant step forward in stem cell application.

In addition to artificial ears, the combination of stem cells and tissue engineering has led to the creation of artificial skin. The skin is the body's first-line of defense against invasion. In the case of extensive burns, scalds, and other traumas, the skin is difficult to repair. If not treated in time, it can allow patho-

human-eared mouse

Cao Yilin

gens such as viruses and bacteria to enter directly into the body, causing infections and other life-threatening problems that also cause great inconvenience
to the patient's daily life. Therefore, to treat this type of patient, skin grafting
is usually used. This involves expanding the skin of the buttocks, abdomen, or
back first, and then transplanting the excess skin to the damaged area to cover
it, resulting in a better treatment effect. If the area of damage is too large and
there is not enough skin to transplant, the skin of other animals, such as pigs or
even fish, is used temporarily to provide a protective covering for a short period
of time. However, these latter treatments are only temporary and usually only
treat the symptoms but not the root cause. Therefore, it is time to consider the
use of artificial skin. However, do not underestimate the skin. Do not think
that it is simply a layer of membrane. From a professional physiological and
anatomical point of view, the skin is an organ with a far more complex structure
than some other organs. From the outermost to the innermost layer of the skin,
it can be divided into the epidermis, the dermis, and the subcutaneous layer.
The subcutaneous layer contains hair, sebaceous glands, erector spinae, arteries,
veins, hair follicles, sweat glands, nerves, and fat cells, among other things. No
matter which layer or small tissue is missing, the function of the skin will be
affected. Up to now, domestic and international scholars have been able to synthesize a small piece of skin in vitro using skin stem cells, artificial gel scaffolds,
and so on. However, due to the incomplete function and insufficient area of the

synthesized skin, there is still a long way to go before it can be applied to the actual treatment of injuries.

As the study progressed and the technology evolved, researchers were not able to simply stack cells and materials to control the formation of precise structures. Therefore, 3D printing technology was applied to the field of stem cells, leading to the formation of an extremely new discipline known as bioprinting. 3D printing began in 1984 when Charles Hull developed the technique of creating dioramas. He constructed many dioramas using resin as a raw material. Twenty years later, the American Thomas Boland had the bright idea to build the first prototype for organ printing using cells as the raw material for inkjet printing. The beginning of the twenty-first century witnessed a boom in bioprinting-based research and development. Every year or two, the world's top scientific journals publish the latest reports from researchers around the world who have printed various animal organs, such as hearts, livers, vessels, and so on, using different gels, cellular matrix materials, and stem cells or other types of cells, either single cells or more complex hybrids, under precise control. Of course, at the moment, all of this is just the form of the organ in the laboratory and is a long way from the size of the actual organ, let alone its function, so it is still a long way from practical application. But in any case, this is a field that is only just beginning. As the technology matures and there is a strong demand market, we are confident that we will be able to benefit from the fruits of this technology within this century.

Ancestors of the Cell Family

$\mathcal{T}$he development of science and technology is often accompanied by controversy, always fluctuating at different stages of history. Among the many cellular research, embryonic stem cells play such a role. They start from a bright light, go through the abyss, come back from the dead, and finally retire from the stage of history.

The reason why embryonic stem cells are so magical is that, as the most primitive cells in development, they have the aura of being able to differentiate into all the cell types of the adult individual. Embryonic stem cells offer new hope for many diseases that cannot be treated or cured by existing methods and for which cellular therapies are theoretically possible but for which there is a lack of stem cells. In 1981, British scientist Martin Evans and American scientist Gail Martin, who had been a postdoctoral fellow in his laboratory, independently reported that embryonic stem cells had been extracted and successfully cultured in mice. Evans was awarded the 2007 Nobel Prize in Physiology or Medicine for being the first to isolate mammalian embryonic stem cells.

Martin Evans

Evans was born on New Year's Day, 1941, in the town of Stroud, England, during the middle of the Second World War when the town was being bombed by Germany. His father's factory was destroyed, and the family was displaced, leaving him with a childhood of hunger and cold. After the war, his father often took him along to work on jobs such as assembling electric lights and repairing generators, giving him strong practical skills from an early age. His first independent experiment in life was to mix sandy cement with water, but it ended in failure. The main reason was that he added too much water, which became a lasting memory for him. That's more advanced than Chinese kids in the 1970s and 1980s who mainly played with mud, but the principle is still the same. By the time he was old enough to go to school, he had become a sickly child with acute appendicitis, mumps, and various infections. He was admitted to the hospital from time to time, leaving him with poor physical fitness but also with the habit of reading. In junior high school, he was more interested in chemistry and physics. He often got excited to hear some technical vocabulary in class. However, in order to gain credits, he was forced to take biology. It was not until he arrived at University of Cambridge that he became interested in botany and genetics, thanks to the relaxed learning atmosphere and the influence of masters from different fields. When he was about to graduate, he missed the entrance exam for graduate school due to illness and could only find a position as a research assistant in the laboratory. Fortunately, his supervisor in the lab encouraged students and staff to use their own initiative, and he felt very relaxed despite not being able to work as a research assistant. It also made him take a lot of detours and step on a lot of potholes in his later doctoral career.

As the saying goes, failure is the mother of success, and these have become valuable lessons in his life. This approach to learning and research is also the principle of his future supervision of students. At the time, he was about to start researching how genetic information "travels" from the nucleus to the cytoplasm of the cell and plays a role. However, the problem was that he either could not get enough cells or he did not know the specific direction of his research. On the advice of a friend, he decided to start with teratomas in mice and applied for a faculty position in the Department of Genetics at the University of Cambridge. He used the stem cells in teratomas as a future research direction. This is a prestigious unit that was thought to be hopeless, and then he managed to get in as a replacement because someone dropped out. This lucky opportunity led to the glory of the rest of his life when, with the help of Matt Kaufman at Cambridge, he succeeded in isolating mouse embryonic stem cells.

Since then, other researchers around the world have obtained embryonic stem cells from other animal sources. However, research into human embryonic stem cells, particularly in vitro cultures, has languished. It wasn't until the late 1990s that success was achieved, thanks to the overt and covert efforts of American James Thomson.

human
embryonic stem cell

James Thomson

Thomson was born on December 20, 1958, in Chicago, the US. Although he studied biophysics as an undergraduate, he double-majored in veterinary medicine and molecular biology during his doctoral studies. After three years of research on in vitro fertilization and embryo experimentation at the National Primate Research Center, he joined the University of Wisconsin-Madison as a resident in veterinary pathology at the age of 33. From there, inspired by his work with mouse embryonic stem cells, he realized that the next step in primate research was already a big one and that he was in a unique position to obtain the raw materials. He cut up a meatball containing embryonic stem cells from a rhesus monkey and, using a special biological technique, digested and disrupted this meatball. Then, he adsorbed the non-embryonic stem cells in the mixed cells onto a Petri dish immobilized with a specific antibody. Finally, he killed these cells, leaving the embryonic stem cells behind. This is a very clever method of reverse screening, removing the unwanted and leaving the wanted. He then transferred these cells to a special culture system containing growth factors and feeder cells from mouse skin cells, which provide a continuous supply of nutrients. After six months of continuous cultivation, in August 1995, the first non-human primate embryonic stem cell line was obtained that could continuously and stably maintain the embryonic state and had the ability to differentiate. These results further encouraged Thomson to pursue the isolation of human embryonic stem cells and the establishment of cell lines. However, from the outset, this research went beyond the scope of a scientific study and left him spinning like a gyroscope, solving one unscientific problem after another.

In the same year, the US Congress passed the famous "Dickey-Wicker Amendment," which prohibited the use of government funds for any research involving the destruction of human embryos. Then, as now, university research institutions, both nationally and internationally, still rely on government grants for their major scientific research funding. If government coffers were to take a bite out of stem cell research, it would undoubtedly make the path of research in this field a thorny one. It was at this time that a private biological company in the United States specializing in aging research and development recognized the potential of embryonic stem cells and was prepared to provide a large endowment to help Thomson continue his research into human embryonic stem cells. With the money, Thomson's next problem was to find a new laboratory. The original laboratory was not available because both the laboratory site and the various equipment in the laboratory had been purchased with government funds. To avoid any conflict of interest, he found an abandoned room in the hospital, bought back some almost obsolete ultra-clean tables from the market to set up a basic cell culture room, and took matters into his own hands to start growing his dream cells. However, the problem always seems to be with him: conducting experiments involving humans must be reviewed by the university's ethics committee and can only be done if the review is approved. Otherwise, it is against ethics. He also knew that this was the eye of the storm of current political opinions. To find out where the minefields were, he consulted his colleagues at the university's Law School, hoping to get help from there. After two years of discussion following the submission of the application report, the committee finally approved Thomson's study. Suffice it to say that the school was also at great risk at this point. If the government vented its anger on the university's other government-funded projects because of this resolution, the damage would be enormous. So even though Thomson had met all the conditions, he still had to hide in that little abandoned room and do what he wanted to do, not daring to make any noise.

To obtain human embryos for his experiments, Thomason on one hand collected surplus embryos from in vitro fertilization at small clinics near the school. On the other hand, he also obtained frozen embryos from a gynecologist at the Rambam Medical Center in Israel. Using the same extraction and culture techniques as was used on the monkeys, he processed a total of 14 embryos. He found cells that resembled the shape of embryonic stem cells, and five of them were able to grow continuously. This made him realize that these were the cells of his dreams. In order to verify that these cells were indeed embryonic stem cells, he first treated them with a dye that can only be used to stain embryonic stem cells. He then had a nurse who passed by the door of his laboratory come in to see the results of his staining. He was thrilled when she told him it was blue, and made the nurse the first person after him to watch human embryonic stem cells grow in a culture dish. Immediately afterward, Thomson performed experiments on the cells, such as teratomas in vivo. After further determining the stem cell properties of the cells, he officially launched the study in November 1998 with the aim of developing them into embryonic stem cells. The results were published in the journal *Science*.

Thomson is an introverted scientist, but as the first person to obtain a human embryonic stem cell line, he still had to move from behind the scenes to the front of the stage. This change of position did not come with applause and flowers, but rather with questioning and opposition from his peers and political and civil ethics organizations. Moreover, as the wave became higher, he was in the center of public opinion more than once. First, once Thomson had succeeded in creating human embryonic stem cells, the university immediately declared that it owned the cells. It not only charged a high fee to the research groups that had purchased them but also forced researchers to sign a patent assignment for new discoveries based on these cells. Why do individual research results end up belonging to the school? It started with a well-known alumni research foundation at the University of Wisconsin-Madison. It approached Thomson at the end of the private company's financial support and signed a series of industry-transformation agreements with him, who was too obsessed with research to be interested in future profits. As a result, ultimate ownership of the cells passed to the university through the foundation. Triggered by these unfair arrangements, scientists who wanted to conduct research using human embryonic stem cells around the world were furious and blamed Thomson, making him a target.

The opposition to Thomson's study mainly came from pro-life activists, who believe that obtaining a person's embryonic stem cells would be tantamount to killing a life. They oppose this in self-righteousness, believing that they are defending the truth on the basis of their reverence for life. However, the facts of science are not what they think. The procurement of embryonic stem cells occurs only in the early stage after the union of sperm and egg. The cells, which are still in the early stages of development and have not yet formed tissues, are cultured in a Petri dish using a special culture system for expansion.

Eventually, an unlimited number of cell lines can be passed on to the next generation. There is no real sense of life in the whole process, especially no generation of consciousness, so it cannot be seen as the strangulation of life at all. But for most of the opponents, even if you explained it a hundred times, it would be like preaching to deaf ears. In the summer of 2001, arguments and objections led to hearings before the highest administrative body in the United States. After much debate, bickering, and lobbying between liberals and conservatives, George W. Bush, the recently inaugurated president of the United States of America who was famous for his conservative political position, finally passed a ban on human embryonic stem cells, which prohibited the use of federal funds for any research that would lead to the creation of new human embryonic stem cells. It seems that a new thing that had just been born was quickly snuffed out. However, as a politician, the president's statement was seen as extremely clever and left room for maneuver. While it was not possible to create new stem cell lines from human embryos, scientific research could be carried out using mature human embryonic stem cells that were already established and in existence at the time.

None of this has deterred scientists from seeking the truth. After all, starting research on human embryonic stem cells is essential for the future development of regenerative medicine. Given the differences between cells and tissues of animal and human origin, it is not possible to extrapolate results directly to humans simply by studying mice or monkeys. The main restrictions imposed by the US federal government included that no federal funds should be used; no new embryonic stem cells should be derived from women; and the use of cell lines already established in vitro was allowed, but they could not be cultured for more than 14 days. It set a red line, and existing studies show that more than two weeks of development produces a wide range of tissues. This rule, therefore, was strictly obeyed by scientists until 2021. In general, there is room for further research on human embryonic stem cells. But it is not enough just to use the cell lines that are already available. For one thing, many scientists have been impressed by the Bush administration's announcement that 60 mature human embryonic stem cell lines have been established. It is very ridiculous because they have no idea where and how to get these 60 so-called cell lines, and these cells alone do not fully satisfy the needs of research. That is why a group of stem cell scientists set out to seek other approaches, and did not stop until they had achieved their goal. Soon after, thanks to the efforts of a group of dedicated individuals, including Hollywood stars, local politicians, and scientists who supported stem cell research, Proposition 71 was passed in California, which supported stem cell research, and raised money through donations from the general public, financial supports from the rich and famous, and the issuance of brand-new bonds, which led to the creation of the California Institute for Regenerative Medicine and recruited a group of dedicated individuals to continue research on human embryonic stem cells. The seeds of hope for human regeneration were planted and grew. In addition, the Howard Hughes Research Foundation in the United States, as the largest privately funded medical research fund, was not subject to the ban and continued to fund Thomson's research with a substantial amount of money. It was these unquenchable sparks that later became the prairie fire. It has to be said that in the general lack of consensus in politics and public opinion, there will always be a group of people in the scientific community who stick to the truth, defy the odds and work for the progress of humankind.

Scientific progress is usually independent of politics, but it is often both constrained and benefited by politics as well, as human embryonic stem cell research has so well demonstrated. Although Bush's administration banned such research, the subsequent administration of Barack Obama quickly legalized and promoted it, allowing it to flourish. Interestingly, resistance is often the impetus for development. Because of the Bush ban, while some people were able to rattle their sabers, there was another group of people who were trying to figure out how to think outside the box and use other cells instead of embryonic stem cells, resulting in the epochal-induced pluripotent stem cells, which we'll discuss in more detail in the next chapter.

Although the lifting of the ban on human embryonic stem cell research has opened a springtime subway to the dream of regeneration, people need to be cautious about operations that may affect future generations, and they need to be in awe of themselves. With such a consensus, late-stage human embryonic stem cell research has been largely confined to ethically acceptable limits, especially in basic research. However, two related studies in China have once again thrown the issue into the eye of the storm.

On April 18, 2015, Huang Junjiu of Sun Yat-Sen University in China published the world's first paper on gene editing of human embryonic stem cells, drawing overwhelming condemnation from the scientific community. However, because the research used fertilized eggs discarded from hospitals and containing multiple sperm cells, which theoretically could not successfully develop into babies, there were no real ethical or moral issues. Huang and his team hence also received supports from some other scientists and the research papers were eventually published in the journal *Protein & Cell*. As the researcher himself discusses in his paper, although the editing was done on genes in human embryonic stem cells associated with thalassemia, a blood disorder that is still common and potentially fatal in the south of the country, the success rate of the editing was not 100%, even with the most advanced gene-editing technology. Given the safety factor and pressure from international colleagues in the industry, Huang did not pursue this research further. Interestingly, the incident was selected by the journal *Nature* as one of the top ten scientific events of the year. It was also the year that the International Summit on Human Genome Editing was organized and produced the first report on the science, ethics and governance of human gene editing, recognizing the feasibility and limitations of human embryo gene editing. Soon after, reports of human embryonic stem cell gene editing began to pour in, seemingly out of control. On the one hand, people were booing, and on the other hand, people were doing what they wanted to do, even though the research was largely limited to basic in vitro research and did not break the bone. So scientific research sometimes turns a blind eye when it's not clear. It is this indulgence and the failure of regulation to keep up with the pace of progress that leads to major incidents.

On November 26, 2018, five days before International AIDS Day, He Jiankui of the Southern University of Science and Technology suddenly announced through the media that he and his team had successfully performed gene editing on human embryonic stem cells. They modified the genes that are susceptible to HIV infection and placed the edited cells back into the mother's womb. As a result, the world's first artificially and intentionally gene-edited human beings were born, a pair of twins named Nana and Lulu. He Jiankui claimed that they had a natural defense against AIDS. For a while, one stone created a thousand ripples, and there was a global outcry. Without in vitro testing, and with social, ethical, scientific, and legal consensus still far from being reached, He Jiankui and his team proceeded directly to edit a person's own genetic material. The changes resulting from this modification will henceforth be integrated into the genetic pool of the entire human race and have the potential to continue in perpetuity. No one knows what will happen to the twins or their descendants in the future. After all, the child has been born and time cannot be reversed.

This time around, human embryonic stem cells are once again at the center of debate in politics, science, media, and general public. The day after his announcement, the Second International Summit on Human Genome Editing in Hong Kong, China, canceled almost all scheduled discussion topics and reports and held a round-the-clock discussion on the incident. Only this time, it was not praise but a crusade. However, He Jiankui himself remained firm and confident, responding to the accusations, "For their sake, we are willing to accept the accusations" and "I firmly believe that ethics will be on our side." Nevertheless, He Jiankui was investigated for violating human ethics, had his activities restricted, and was eventually imprisoned. Two years later, China also included the post-implantation of human embryos in the Criminal Law, providing a national legal response to regulate research in this direction.

The purpose of human embryonic stem cell research is unquestionable if it is for the treatment of disease, but it is still highly controversial whether it is possible to target human populations for modification to improve some currently seemingly superior trait, as is the case with animal or plant breeding. Looking back at the case of He Jiankui, there are at least three scientific problems. First, inadequate scientific ethics and non-compliance in operational processes; second, the wrong gene was selected for correction, although the gene is an important receptor gene for HIV infection of blood cells, the absence of the gene can lead to other health problems at the same time; third, the gene-editing technology used is still immature and there are huge risks and loopholes. Because of these problems, He Jiankui's punishment is not excessive. However, with the completion of ethics, the maturity of the technology, and the regulation in place, embryonic stem cells may be modified to help people recover from genetic diseases, benefiting the human race in the future.

Compared with the above major events in human embryonic stem cells, the research and application of animal embryonic stem cells have been much smoother and have produced many influential results. One of the most notable contributions has been the acquisition of various types of genetically modified animals. Scientists have used mouse embryonic stem cells as research subjects and inserted a gene that can produce green fluorescent protein. This resulted in mice that can emit green light when exposed to ultraviolet light. It reminds me of the scene in the film *Avatar*, where both plants and animals glow in the eerie night light. And this is just a piece of cake. The real contribution lies in the fact that it is possible to make changes to any gene in embryonic stem cells, either to enhance or weaken the function of such genes. Scientists are able to do this and then study the animals after birth, especially mice, to simulate a wide range of human diseases. This provides a model for the study of human diseases. The Jackson Laboratory in the United States has become the world's largest base for genetically modified animals, serving tens of thousands of laboratories around the world each year. It all began with the unintentional action of a postdoctoral researcher. This is another story of "a watched flower never blooms, but an untended willow grows."

Rudolf Jaenisch was born in Germany in 1942. Coming from a medical family, with both his grandfather and father being internists, it was only logical that he chose to study medicine at university. But after a brief exposure to anatomy and physiology courses, he realized that he had no interest in them at all and was easily distracted in class. This was not an option, so he had to change. He decided to attend classes while conducting experiments at the Max Planck Institute for Biochemistry. It was also at this time that he slowly realized that his life's work would be to carry out scientific experiments and study the unknown, rather than to learn what was contained in those medical books. It was at this time that phage, an important research tool in modern molecular biology, became the first experimental object he came into contact with. After graduating, he decided to continue his postdoctoral training in the US. After contacting a circle of possible mentors, he decided to study under Arnold Levine at Princeton University. The main reason for this choice was that Levine's research background was also in phages and included some genetics, particularly on how viruses induce tumor production. He found this topic very interesting. The first project he took on was to use monkey vacuolar viruses to infect animal cells and then test whether or not they produced tumors. This is a serious subject because tumor production mainly occurs in adulthood or old age. Therefore, generally injecting viruses into animals of that age is the right way to conduct research. Otherwise, it is just nonsense. But once his mentor went on holiday to Europe and was away from the lab for a long time, he was left to amuse himself. This was great because "when there is no tiger in the mountain, the monkey becomes the king." He ordered the graduate students in the lab to inject the virus into mouse embryos. His idea was that if the virus was injected locally in an adult animal, it could only lead to the production of one type of tumor. If it was infected at the embryonic stage, theoretically, every cell would turn into a tumor cell, but the results were not what he expected. As a freshly minted postdoctoral researcher, he did not have enough knowledge to understand the experimental results. At this point, with his mentor still abroad, he turned to Professor Beatrice Mintz, a developmental biologist in Philadelphia, for help. She advised him to repeat these experiments and confirm the phenomena before discussing them, otherwise they would make no sense. With her help, Jaenisch infected the embryos with the virus and then transplanted them into another surrogate mouse. This resulted in the mice being fully infected with the virus. However, the mice still didn't develop tumors, and once again the results of the experiment were not as expected. He reached the impasse and deadlock that every scientist often encounters.

Rudolf Jaenisch

At this time, he was fortunate enough to start his own independent research group at the Salk Institute in California. This new environment always
brings new ideas. At the suggestion of his new colleagues, the use of so-called
thermal methods, in which viruses are labeled with radioactive elements, made
it possible to analyze whether the genetic material of the virus had entered the
genetic material of the embryonic cells. Accordingly, it was possible to analyze
the probable reasons for the failure of the experiment. After several experiments
with the thermal method, Jaenisch was 100% certain that the virus had indeed
entered the embryonic cells and was present in the cells of the mice that were
born. In this way, he had inadvertently constructed the world's first genetically
modified animal. However, at this point, he still had not solved the problem
of the relationship between the virus and tumors, nor why this virus did not
induce tumor formation. It was this inadvertent work that won him numerous
awards and ushered in an era of transgenic functionality. Of course, building
on his foundation, subsequent researchers, in order to increase efficiency and
precision, no longer injected viruses carrying genes for different traits directly
into embryos, but instead directly manipulated embryonic stem cells in culture
flasks and then transplanted the gene-edited embryonic stem cells into recipient animals to produce transgenic mice, transgenic fish, and transgenic pigs,
among others.

The creation and acquisition of these genetically modified animals, in addition to basic scientific research on applications, can also be used for food as

well as for the treatment of disease. In the field of regenerative medicine, cell biologists have considered using stem cells for therapeutic purposes, while others have considered directly replacing lost organs through organ transplantation. In the field of regenerative medicine, cell biologists think in terms of stem cell therapy, while others think in terms of direct organ transplantation to replace lost organs. However, there are far fewer human organ donors than people who need them, so the possibility of using animal organs for transplantation is as old a research topic as the early days of transfusing animal blood cells into the human body. As the number of experiments increased, researchers came to two conclusions. First, of all the animals tested so far, pig organs are the closest to humans in terms of shape, size, and function; and second, the main reason for the failure of animal organ transplants into humans is allergic reactions. For this reason, researchers have been working hard in these two directions. They are using pig embryonic stem cells as the research object and deleting the main genetic material that causes allergic reactions, the galactosidase gene. This produces transgenic pigs, theoretically making their organs suitable for human transplantation without causing serious allergic reactions. The FDA-approved pig is a galactose-safe pig developed by Revivocor. Its organs, such as the heart and liver, can be used for organ transplants in patients with heart or liver failure. The leftover pork can be consumed as normal, making it a two-for-one deal. Of course, there is still a lot of clinical and safety research to be done before we can say for sure that this genetically modified pig is as good as we want it to be. One of the key issues, for example, is that the pig's body contains animal-derived viruses that, in theory, cannot infect humans. However, the question whether or not unexpected diseases can occur when organs are transplanted into the human body also urgently needs to be addressed.

In China, Academician Zhou Qi's team at the Institute of Zoology, Chinese Academy of Sciences, is conducting a human class of embryonic stem cells for the treatment of related diseases. At present, their main idea is to turn embryonic stem cells into dopaminergic neural progenitor cells and then use stereotactic injection technology to transplant them into the diseased brain area of Parkinson's disease. This is based on the completion of a series of preliminary animal experiments. They have also started to move toward clinical trials. In 2017, relying on the First Hospital Affiliated to Zhengzhou University, they slowly injected four million cells into the brains of Parkinson's patients, launching the world's first clinical treatment trial. We'll have to wait and see what the final result will be, whether it has no effect, delays the disease, or improves it. Meanwhile, other countries around the world have conducted clinical trials for age-related macular degeneration, diabetes, and osteoarthritis using human embryonic stem cell-derived cells, including retinal pigment epithelial cells, pancreatic islet cells, and MSCs. Among them, Douglas Melton of Harvard University in the United States is a leading figure in the treatment of diabetes using islet cells derived from the differentiation of embryonic stem cells. He started out as a researcher on frog development, but it was his son's long suffering from the disease that led him to turn to human embryonic stem cell research. He firmly believed that this was the only hope for a cure. He had followed Thomson's work from the beginning and had experienced the highs and lows of the field from beginning to end. His mood had changed from one of blissful excitement to one of heartbreak, and now to one of serenity. After all, any major scientific and technological discovery to application, will always not be smooth sailing, as little as a few decades, more than a hundred years, are very normal. As long as there is hope, no matter how bumpy the road, we will take solid steps toward our destination. This is the mood not only of Melton, but also of many obscure workers in science and technology. The saying of "not being happy for things and not being sad for yourself" could not be more appropriate here.

Cells in a Time Machine

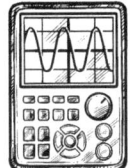

$\mathcal{T}$ he fact that pluripotent stem cells have such attractive application prospects and embryonic stem cells are so loved and hated means that people have to try to find other ways out. Otherwise, the door of regenerative medicine can only be pushed open when there is a gap. Then it is slammed shut tight, letting the sunshine in but never allowing the silhouette to come and go.

At the time, there was a postdoctoral fellow from Japan named Yamanaka Shinya at the Gladstone Institute of Cardiovascular Disease in San Francisco, California. His main research was to find out the mechanism by which a protein assigned to him by his mentor worked to lower blood lipid levels, theoretically preventing atherosclerosis. To carry out his research, he first created a transgenic mouse to study the state of the mouse when the gene encoding this protein was fully activated. The expected result was that such mice would have a hard time developing the aforementioned disease. However, before these transgenic mice could produce the phenomenon they wanted to observe, they all developed tumors. Obviously, this was a very bad outcome. Although his mentor was also disappointed with the results of his experiment, he was encouraged to continue the project and find out why this gene caused the tumors to develop. Following this lead, he found another gene associated with this gene. At this point, he needed to use embryonic stem cells to delete the target gene and create transgenic mice. However, he had no experience in culturing stem cells. Fortunately, his institute was full of talented people, and with the help of other experts in his group, he quickly mastered all the skills involved in isolating and culturing mouse embryonic stem cells and genetic manipulation. It was at this point that he formally entered the stem cell field and laid the foundations for the future embryonic stem cell revolution. But it wasn't long before his wife returned home with their two daughters to look after their eldest daughter at school, and he had no choice but to pack his bags and return to Osaka, Japan, from San Francisco six months later.

Osaka is located in the west-central region of Japan, second only to Tokyo in terms of population. It is about 400 kilometers from Tokyo and is surrounded by mountains on three sides and water on the other. The city has four distinct seasons and a pleasant climate with warm winters and little snow. Nara is a half-hour drive to the east, and Kyoto is about 30 kilometers to the northeast. Coincidentally, Osaka and the US city of San Francisco established a sister city relationship in the 1950s, which seemed destined to be a match for Yamanaka Shinya. Yamanaka Shinya was born in the same city on September 4, 1962, where his parents ran a small factory inherited from his grandfather that made parts for chainsaws. His mother ran the factory and took care of everything, while his father, a well-known local engineer, was very good at designing and manufacturing new products. This had a subtle influence on him as a child, growing up tinkering with clocks and radios that were often taken apart but could not be put back together. His dream was to become an engineer when he grew up. In primary school, he enjoyed reading science magazines and visiting science labs. He once caused a fire by knocking over an alcohol lamp and was scolded by his mother. When he was in junior high school, his father encouraged him to take up judo as a way of strengthening his slender body. Judo is a competitive sport that originated in Chinese martial arts but was developed in Japan. In this sport, two people fight each other. As a result, he was either bruised or broken. Eventually, on his father's advice, he enrolled in college to study medicine and gave up judo shortly thereafter.

After graduating at the age of 25, like all medical graduates, he worked in a hospital and began the first stage of his medical career as a junior doctor. During this time, he experienced great joys and sorrows in his life. First, he married a schoolmate he had known since junior high school, and then his father, who had long suffered from diabetes and hepatitis, passed away. In addition, he found himself to be a clumsy surgeon who usually took two hours to perform an operation that would take other doctors only 30 minutes. As a result, his boss was really unsatisfied with him, which made him lose his interest and confidence in being a doctor. But was his own compassion and thinking about the suffering of patients that led him to eventually abandon the practice of medicine and pursue basic medical research. An operation may save one person, but not more. A breakthrough in basic research can fundamentally help more people. So he decided to pursue a PhD in pharmacology. His PhD study had a significant impact on his lifelong research career, particularly his supervisors, who taught him how to read the literature, design experiments, perform experimental procedures, and analyze data. After obtaining his PhD, he continued his postdoctoral research at the Gladstone Institute of Cardiovascular Disease in San Francisco, the US, as mentioned above.

After completing his postdoctoral work and returning to Japan, he secured a position as an assistant professor in the Department of Pharmacy at Osaka City University. Fortunately, his postdoctoral supervisor allowed him to bring the genetically modified mice he had studied in the US back to Japan. The head of department at his new institution did not restrict his research direction, allowing him to continue the topics he had not completed during his postdoctoral period. It was at this point that he discovered that the loss of the gene he had previously focused on was fatal in mice. The reason for this was that it prevented the normal differentiation and development of embryonic stem cells. The next step was to investigate how embryonic stem cells maintain their own identity. As a faculty member in the Department of Pharmaceutical Sciences, most faculty members were involved in research projects with clinical translational potential. His research was so basic that he felt isolated with few people to communicate with and receive positive advice from. Lucky enough, Nara University of Science and Technology agreed to offer him an associate professor position with a substantial grant to help him start his research. The only difference was that he would have to help run an animal experimentation center. However, it was this seemingly onerous position that led to his major discovery that would transform the field of regenerative medicine.

At the age of 37, Yamanaka Shinya has finally set up his own independent research group with a leader position, a more generous research grant, and a relatively good previous research record that allowed him to take a breather and do some challenging research. To this end, he found a guy who had just published a nice paper and didn't have to worry about getting new results published, at least in the short term, to test his crazy ideas. The idea was this: since embryonic stem cells are so important and difficult to obtain, could they be produced artificially? According to the results of his previous research and the conclusions already reported by others, all cells have the same nucleus, which contains the same genetic material. The key factor in determining the different personalities of each cell lies in the different proteins that are expressed. He has identified a number of key proteins, and if these key proteins are introduced into adult cells, it is theoretically possible to make these cells become embryonic stem cells that everyone wants. In the stem cell field, there was an old man named Waddington in 1957 who put forward a very vivid analogy. He said that cell development is like a ball rolling down from the top of a hill, through different divisions, and flowing into different foothills, eventually becoming different types and functions of cells. If the ball is pushed back to the top of the hill from the foothills, it can regain the ability to roll down again. Of course, this was all speculation, hypothesis, and no one had ever tried it. And what he was trying to do now was to test the possibility of this hypothesis. Another thing that made him dare to try this bold idea was that at that time, the whole international community of his peers were all scrambling to study the differentiation of embryonic stem cells, which in today's buzzword is a kind of severe involution. If he wanted to stand out, he had to do the opposite.

In order to find the motivational factors that make the ball roll uphill, he first needed to find out why embryonic stem cells have pluripotent properties while other adult stem cells don't. He also needed to determine what the differences are between them and whether there are key factors that control the special mission of embryonic stem cells. All he had to do was find those key factors and discover the connections between them. Just like a detective solving a case, when you intervene in a gang case, it's hard to root out the gang in one fell swoop. It's usually a matter of delineating the range of suspects, checking them out one by one, and then finding a breakthrough and following the clues. Although he had several candidate factors, these were not enough. Based on his and his students' brainstorming, by summarizing, comparing, and analyzing data published by others and in publicly available databases, they quickly identified more than 100 prime "suspects" for maintaining the properties of embryonic stem cells. The final 24 were identified as the factors that would actually be manipulated.

Good research always needs good ideas, but it also needs to make good use of the resources around it in order to flourish. Otherwise, ideas will always be just ideas, like unattainable illusions. As a laboratory animal research center, they have a variety of mice resources, including a more special mouse. He entered the new unit and successfully constructed the first transgenic mouse, but because of the phenotype, he did not achieve the expected results, so basically, it is considered a failure of the subject to produce useless mice. Because of the commemorative significance, he was reluctant to discard it. At that moment, he had a flash of inspiration about how to use this mouse. Since there was a gene in this mouse that was specifically expressed only in embryonic stem cells if this gene was followed by a drug-resistant gene, only the embryonic stem cell would survive in a pile of mixed cells under the conditions of the killing drug, while all the other cells would die. At first, they introduced the 24 gene candidates into the starting cells one by one. However, no cells survived the screening drug. Later, they refined their strategy and carried out an extremely painstaking and correct experimental design. They combined the genes and put them together in the somatic cells of this group of mice. Through this seemingly endless series of daily experiments, they discovered for the first time cells that looked very much like embryonic stem cells and could survive. After a brief moment of excitement, the scientist's mind took over. In response to this discovery, Yamanaka Shinya and his team first analyzed the genes that led to the success of the experiment. Then, they repeated the process by introducing these genes into the cells again and again. For a scientific discovery, whether macro or micro, only those results that can be repeated have value. Their iterations validated their initial findings and narrowed down the effectors from the 24 who were earlier candidates to the four who played a key role.

In 2006, they published the results and named the artificially induced embryonic stem cells as induced pluripotent stem cells, which many people now refer to simply as iPS cells. Yamanaka Shinya didn't advertise this result everywhere, but it quickly attracted a lot of interest and attention from his colleagues. Great discoveries, as soon as they show the slightest glimmer of light, grab the limelight and compete for the fruits that follow. Given the success of artificial embryonic stem cells in mice, the next step was to use the same strategy to test whether artificial embryonic stem cells of human origin could be obtained. The group he led soon competed with Thomson's research team in the United States, and at the same time in the following year, both reported success in obtaining artificial embryonic stem cells of human origin. In fact, from the day he received the mouse results, he began to prepare for human experiments. Given the limitations of the conditions in Nara, he was promoted to professor in 2004 and moved to the Institute of Frontier Medical Sciences at Kyoto University. At the time, it was the only institute in Japan that could culture human embryonic stem cells. Proudly, the first person to derive the iPS cell in the US team is a Chinese scholar named Yu Junying. She had graduated from Peking University and was doing her postdoctoral training in Thomson's group. Immediately afterward, a whirlwind of artificial embryonic stem cell research took place around the world. In China, Pei Duanqing's group at the Guangzhou Institute of Health Research of the Chinese Academy of Sciences was the first to confirm the reproducibility of the research. Xiao Lei's group at the Shanghai Institutes for Life Sciences of the Chinese Academy of

Sciences was the first to obtain induced pluripotent stem cells from rats. Zhou Qi's group at the Institute of Zoological Research of the Chinese Academy of Sciences in Beijing, China, collaborated with Zeng Fanyi at Shanghai Jiao Tong University, China, was the first to show that these induced mouse pluripotent stem cells could completely replace embryonic stem cells and develop into a normal mouse. After that, it can be said that nine out of ten stem cell research teams in the world have been working on artificial embryonic stem cells. Some are studying the mechanism, while others are studying how to use them further. Soon after, Yamanaka Shinya was awarded the Nobel Prize in Physiology or Medicine in 2012. The person who shared the prize with him was the British, Sir Gordon, who will be mentioned in the next chapter. As a result, China has entered the fast lane of stem cell development. Not only is it keeping pace with its international counterparts in this field, but it is also taking the lead in certain directions.

Yamanaka Shinya

iPS cell

As the well-deserved father of iPS, Yamanaka Shinya, with the support of the Japanese government, established the iPS Research and Application Center and recruited many followers to join him in trying to roll this cell, which had been pushed to the top of the mountain, back down to the designated valley according to people's wishes. This way, the direction of the cell can be controlled and the type of cell that we want to use for treatment can be obtained with precision, for example, cardiomyocytes, neuronal cells, blood cells, and so on. However, history moves forward with twists and turns, and technological progress has always been spiraling. iPS has solved the problem of the source of pluripotent stem cells, but if it is to be used in a real sense, especially for the treatment of various diseases, it still has to overcome numerous problems. There are three key issues. First, when all the pluripotent stem cells roll from the top of the hill to the bottom of the hill, if the cells that have not completely rolled down to the bottom of the hill but are still on the slope of the hill or near the top of the hill are accidentally collected together, will there be any safety problems, such as the tumourigenicity of the embryonic stem cells in vivo? Second, how can we precisely achieve the directed reprogramming of pluripotent stem cells to become the cells we want, and the higher the efficiency the better? Third, how can iPS-derived cells be transplanted to a focal site for therapeutic effects? Of course, there are many more questions, and these are the general direction of the problems that scientists urgently need to solve at the moment, and the specifics of different cell types will give rise to many smaller problems.

In 2017, Yamanaka Shinya and his research partners reported the first case of age-related macular degeneration using iPS-derived retinal pigment cells. The disease, which occurs mainly in the elderly, is classified as wet type and dry type. The former, which is mainly due to the production of excess blood vessels, occurs in far greater numbers than the latter, which is due to atrophy. As the disease progresses, it causes irreversible damage to the pigment epithelium in the retina, which can lead to blindness. The traditional treatment options are either to give drugs to inhibit the growth of blood vessels or to use laser surgery to burn off excess blood vessels. However, these methods treat symptoms rather than the disease, and there is still no way to restore damaged cells. Some people have also tried to replace these lost cells by isolating cells directly from the patient's eye, culturing and growing them, and then injecting them back into the eye. However, this method is difficult to promote due to its traumatic nature. Besides, as mentioned earlier, when human embryonic stem cells are available, some people have used their differentiation to obtain pigment epithelial cells and then carried out transplantation therapy. However, there are limitations in the source of embryonic stem cells and matching the cells, making it easy to produce rejection. Therefore, the use of iPS-derived cells for treatment can be considered a good attempt and choice.

iPS cells

They selected two patients, one a 77-year-old woman with lesions in both eyes and the other a 68-year-old man with a lesion in his right eye. The trial started after the ethics committee approved the study and the patients signed an informed consent form. As a Phase I clinical trial, the primary objective was to observe the safety profile while the patients were conditionally treated to improve their vision. To ensure greater safety, the cells isolated from the skin of each of the two patients were first converted into iPS cells using a viral-free method and then induced to differentiate into pigment epithelial cells. These cells were tested for various indicators to ensure that they did not have a tendency to malignant transformation and that they had the function of a normal pigment epithelial cell. The transplantation was not carried out by injecting the dispersed cells directly into the damaged retina, but by seeding around 100,000 cells on a small piece of gel to form a cell membrane that would eventually adhere to the diseased part of the retina to complete the transplantation. It took 13 months and 4 months, respectively, from the start of hospitalization to the completion of the transplant operation. Both patients were followed up on average every two months for two years after the treatment. The results of the follow-up showed that the treatment had an excellent safety profile. Although no significant improvement in the patients' vision was observed, at least the progression of the disease was slowed and did not worsen, demonstrating the great potential of this method.

The creation of iPS cells has made it possible to individualize stem cell therapy, such as to create iPS cells that belong to the patient. This has the obvious advantage of avoiding the need for immunization. In addition to the problem of exclusion, there is also a major disadvantage: the cost of treatment is extremely high, which makes it difficult to use it as a means of promoting it in the future. For this reason, the international community has begun to prepare for the establishment of iPS banks, just as it has for the establishment of blood, cord blood, and bone marrow banks. In China, two such cell banks have been established in Beijing and Shanghai, respectively, based on the implementation of major national R&D programs. It is worth noting that the aforementioned Chinese scientist, the first Chinese creator of human-derived iPS, has also returned to China, established a private iPS bank, and carried out the development of iPS-derived MSCs, dopaminergic progenitor cells, and cardiomyocytes, as well as the treatment of corresponding indications. It can be seen that both international and domestic, national and private companies are focusing on iPS cells. This will surely accelerate their application and bring the fruits of regenerative medicine in the near future.

Influenced and inspired by the iPS storm, since cells at the bottom of a hill can be pushed back to the top, cells at the bottom of one hill could theoretically be pushed directly onto the other, thus transforming one type of cell directly into another type of cell, without having to laboriously push the cell back to the top of the hill and then let it roll back down to the bottom of the hill. Based on these crazy ideas, scientists are trying again and again, night after night, competing to be the first to push different types of cells from one hill to another. For example, taking cells from urine and turning them directly into nerve-like cells or pulp stem cells that can be regenerated into teeth; cutting a bit of tissue from the skin and turning it into blood cells or cardiomyocytes. The creation of artificial blood cells, in particular, could, if successful, put an end to the current practice of blood donation, which is heavily promoted. This could not only revolutionize medical treatment but also change people's lives and influence national policies. This artificial cell research for the future of regenerative medicine offers unlimited imagination. These imaginations are not fantasies, but can really be achieved with our existing scientific and technological means. Everything is just a matter of time.

If Waddington was the first to come up with the idea, it was Robert Davis and others at the University of Washington, the US, who put it into practice and proved experimentally that it worked. In 1987, they found that the action of a single gene was able to push the cells in the skin of mice over the hills in Waddington's model, rolling them directly from one valley to another and turning them into muscle cells. To verify the gene's powerful driving ability, they repeated the experiment with fat cells and kidney cells. Both types of cells could be pushed from different valleys into the valley where the muscle cells were camped. At the time, however, the results of this study attracted no interest, and the article was published as if a small stone had been thrown into a river pond, causing only slight ripples and not even a splash. But Yamanaka Shinya read the article and was inspired by this early article to make his own crazy and reliable attempt. In the fourth year after the publication of Yamanaka's article, someone made an attempt to create a new model for the world. It took the use of one of the four important factors he discovered to drive skin cells directly into blood cells for people to realize the value and importance of that early article. While it is unfortunate that Davis et al. did not end up sharing the Nobel Prize in Physiology or Medicine with Yamanaka Shinya, their pioneering contributions are still very much recognized and appreciated. To be awarded the Nobel Prize is an affirmation; to be recognized and remembered by future generations is an even greater affirmation.

Whether it is pushing the cells to the top of the mountain or from one valley to another, the ultimate goal is to use them to treat clinical patients. In all previous studies, the protocols that have been used to push the cells have basically involved the use of viruses, which has led to safety concerns at a later stage. Admittedly, we have already modified viruses to some extent and domesticated a certain type of virus for our use. However, accidents always happen. As the saying goes, "A dog will jump over the wall if it is nervous, and a rabbit will bite if it is in anger." In order to improve safety and reduce the cost of producing these cells at the same time, scientists are also trying their best to find other boosters. This is the only way to make the most advanced medical technology available to the general public, instead of becoming the exclusive privilege of certain powerful and wealthy people, just as Elon Musk challenged the launch of rockets by improving disposable rockets into reusable ones, greatly reducing the cost and opening the way for everyone to enjoy space travel in the future. In this direction, Chinese scientists have achieved research results that are ahead of other countries. Adopting the compound cocktail method instead of the virus and gene method, which has the advantages of high safety, easy operation, and low cost, they have not only successfully pushed the valley cells in mice and humans to the top of the mountain and obtained chemically induced pluripotent stem cells, but also succeeded in turning skin cells directly into neurons, cardiomyocytes, blood cells, and so on, in both the mouse and human systems. We can imagine how amazing it will be one day in the future when all the technology is mature. When a leukemia patient urgently needs a blood transfusion or a hematopoietic stem cell transplant, just a few cells need to be taken from the skin. They can be thrown into a cocktail of culture flasks mixed with a variety of compounds, and as if by magic, blood can be obtained for treatment. And these are not fantasies; they are happening little by little in the laboratories of our scientists. It's just that ordinary people haven't known it yet.

Scientific experiments cannot afford to be sloppy, and if the steps are too big, just like driving a car too fast, it is easy to have an accident. Looking for a more precise way to change cell fate, Obokata Haruko of Japan's Institute of Physical and Chemical Research claimed to have found a simple method of acid treatment. According to her report, by simply placing the cells to be treated in an acidic culture solution and soaking them in it, and then using a squeezing method to squeeze each cell, these cells could be returned from their position in the valley to the top of the mountain. Amazing as it sounded, not only the method was simple, but the cost was also so low that almost anyone, even an old farmer who worked the land or an auntie who was a square dancer, could do it. If the results were as she described, the contribution to humankind would be absolutely enormous. However, within a few months, while the cheers had not died down yet, almost every other scientist working on this research around the world was complaining that the method could not be duplicated. In the scientific community, research that cannot be repeated is despised. Obokata Haruko, under immense pressure and close supervision, began to repeat the experiment and embarked on the path of self-proof. However, before the results were available, her mentor, Sasai Yoshiki, who had pioneered organoid technology in the field of stem cells and was a strong candidate for the Nobel Prize in Physiology or Medicine, commited suicide in his laboratory in a fit of unbearable humiliation. This became one of the few scandals in the field of stem cell research, known as the "Obokata Haruko Scandal."

Since the destiny of cells can break the limits of natural development, people will naturally think that similar technology can be used in humans to return to childhood. Whether in ancient or modern times, at home or abroad, returning to youth or achieving immortality has always been the goal of people's dreams. In China, after the unification of China in 219 BC, Emperor Qin Shi Huang began to search for the art of immortality. To this end, he successively sent Xu Fu and Lu Sheng with a fleet of 1,000 boys and girls to the three mountains in Bohai Bay, namely Penglai, Fangzhang, and Yingzhou, in search of the elixir of immortality. Not only did he fail to achieve longevity, but the Qin Dynasty (221–207 BC) also fell prematurely. In 2008, the American film *The Curious Case of Benjamin Button* was released, depicting the fantastic life of Benjamin Button. He was already over 80 years old when he was born, but as the years passed and his relatives and friends grew older, he became younger and younger. He changed from an old man to a middle-aged man, then from a little boy to a baby. Finally, he dies in the arms of his aging lover. The story of turning back the clock in the film is, of course, fictional. However, the character himself is based on reality. There is a disease called progeria, or premature aging syndrome. Children with this disease are aging five to ten times faster than normal, and even at a very young age, they look like old people. Not only their appearance but also their body organs show rapid aging. For these people, it is already considered a long life to be able to live into their teens.

Juan Carlos Izpisua Belmonte

Today, in addition to the various health care burdens brought about by the increasing aging population of human society, as well as social and livelihood issues, research into healthy aging has become a scientific discipline. Of course, the purpose of the study is not to be like Qin Shi Huang or Button, but to slow down the aging process so that many of the various diseases that should have begun to appear at the age of 60s or 70s will be delayed until the age of 80s or 90s before they begin to have an effect, in which case, by the time we reach the age of 60s or 70s, we will still look as young and active as we did in our 30s or 40s. To put it in a fashionable and scientific way, youth should be temporarily eternal. Somewhat earlier scientific experiments on monkeys, followed for up to 20 years, have shown that the reduction in metabolic activity caused by dietary restriction is a potentially effective way of slowing aging. In addition, scattered studies have found that a variety of natural products, such as resveratrol in red wine, not only have the ability to accelerate the rolling of cells from the bottom of the hill to the top but also show better anti-aging properties in mice and fruit flies. Spanish scientist Juan Carlos Izpisua Belmonte is an active researcher and pioneer in this field. Using genetic manipulation methods, he obtained transgenic mice with progeria and precisely controlled the expression of the four key factors discovered by Yamanaka Shinya in these mice. He found that not only was the aging of these mice slowed down, but the characteristics of progeria were also reversed to some extent. Additionally, the decline in regenerative ability after various tissue and organ damage associated with aging was well compensated. All of the above research has provided the impetus and direction to continue with the scientific approach to anti-aging. When the day comes that technology is no longer an obstacle and immortality becomes a reality, the various legal and ethical issues that will be raised will be the most debated. The curse of stem cells will be conjured up once again, but this time not for embryonic stem cells, but for iPS.

Cloning from Cells

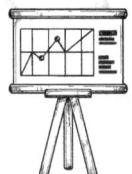

$\mathcal{T}$he word "cloning" is a very sci-fi and fascinating word. Films such as *The Fifth Day* and *Escape from the Clone Islands* have become classic sci-fi films with cloning techniques as the backdrop. In fact, the term clone has been around for more than half a century. The simplest and most straightforward explanation of cloning is that one can become two, and those two are exactly the same. Human cloning is still a dream, if not a reality, but cell cloning has long been achieved. With the gradual maturation of technology, animal cloning has been achieved in many species. One of the best-known animals is Dolly the cloned sheep. A short article published in the journal *Nature* had swept the world like a whirlwind, through the scientific community, the agricultural community, ordinary people, and even the world of entertainment, which has nothing to do with it.

Before we get to Dolly's story, we need to know a little about Sir John Gordon, who pioneered the field. Sir Gordon is a professor at the University of Oxford, England, and has also been knighted by the Queen of Britain. Of course, this was after he became famous. Before that, he was an undoubtedly stupid student. He was born on October 2, 1933, in Frensham, a small village in the south of England dotted with moors and ponds. Although his mother was born a farmer, his father came from a prominent family dating back to the twelfth century, many of whom were government and local officials. But by his grandfather's generation, the family had fallen on hard times and could no longer afford to live in the ancestral mansion, which was burned down during the Second World War. His father dropped out of school at 16 and became an apprentice in a rice trading house. After serving with distinction as a volunteer in one of the world wars, he was discharged from the army and took a clerical job in a bank until he retired. Gordon's childhood was a time of war and hardship when scrimping and saving were the norm in the family, and when the family's past glories were such a fantasy that he didn't know what a banana or an orange was. Perhaps it was these factors that made him clumsy as a child. At the age of eight, during an entrance examination for a local private school, the examiner asked him to draw an orange, but he drew a tree because, in his mind, oranges had to grow on trees, not hang in the air. The examiner tore up his drawing on the spot and told his parents, "Your son has an intellectual developmental disorder and should go to a special school, not here." He had no choice but to go to another school, which fortunately was not so strictly regulated, and he was let loose, spending all day observing insects and plants and collecting eggs and larvae of all kinds of insects, which became a lifelong hobby, even after he had made a name for himself. But learning was never easy for the teenage Gordon, either emotionally or intellectually. At the age of 13, he moved to another boarding school but was still often bullied by older students. To avoid them, he chose to spend his time playing tennis, an unpopular sport at that time that required not much skill, just brute force. On top of this, he did poorly in his elective science and biology courses at school, often finishing at the bottom among the 250 students in his grade. His biology teacher wrote a report to his parents describing him as the worst of the worst. It read, "It has been a disastrous half. His work has been far from satisfactory. His prepared stuff has been badly learned, and several of his test pieces have been torn over; one of such pieces of prepared work scored 2 marks out of a possible 50. His other work has been equally bad, and several times he has been in trouble because he will not listen but will insist on doing his work in his own way. I believe he has ideas about becoming a scientist; on his present showing this is quite ridiculous, if he can't learn simple biological facts he would have no chance of doing the work of a specialist, and it would be a sheer waste of time, both on his part and of those who have to teach him." Gordon has kept this little note to this day, and it sits on his desk at work. But that's not all, the year he went to university he failed due to his poor grades, but luckily his family, through a circle of maneuvering, gave him the opportunity to enter Oxford as a matriculated student, but only if he re-sat his A-level physics, chemistry and biology courses and then took another exam. Fortunately, this time he passed and finally got into Oxford as a matriculated student, majoring in zoology. Turning a hobby into a career couldn't have been more enjoyable for anyone, and it was no different for Gordon. So when he left college, he wanted to do a PhD in zoology. To this end, he spent his days wandering around the school grounds with a trap net and actually managed to catch a new species of insect that had never been seen in the UK before, but still, his application for a PhD in entomology was rejected. Fortunately, he is always lucky and always comes through at the right time. This time, developmental biologist Professor Michael Fischberg offered him an olive branch.

Sir John Gordon

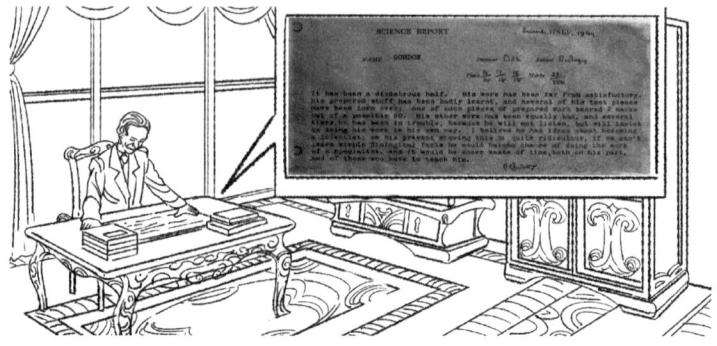

Perhaps Gordon's heart is bigger, or perhaps his self-esteem is higher than that of his peers. In short, Gordon chose to make basic scientific research his life's work in his later studies and work. In the early days of his scientific career, the research craze at the time was the observation of early animal development. However, even at a prestigious institution like the University of Oxford, there was not much research funding and research equipment. Gordon, as a fledgling young man, was no better off. All of his research equipment consisted only of a microscope. Compared with today's research facilities, it was way too simple. In modern biology laboratories, microscopes are already the most basic equipment and are even negligible. Only electron microscopes, two-photon microscopes, multicolor flow cytometers, mass spectrometers, single-cell sequencers, and multifunctional high-throughput drug screens are considered scientific equipment. But life must go on, science must go on. Unlike the old saying in China, in our real life, you can't give up crafting the porcelain simply because you do not have a drill made of diamond. As long as there is a force driving, people can always create conditions in the absence of conditions. For Gordon, it was the use of such a simple microscope that began his great scientific journey.

According to the basic principles of development, it is generally accepted that once cells have developed from early embryonic stem cells into adult cells, they must have lost the characteristics of embryonic stem cells. This was the logical consensus at the time. From a technical point of view, although every cell has a nucleus containing the genetic material, there is a significant difference between an embryonic cell, which has the ability to develop into a fully-fledged individual, and an adult cell, which can only carry out the functions of its own local organization. To challenge this supposedly authoritative consensus, many have attempted to disprove it through experimentation. The simplest experiment is to transplant nuclei from adult cells into egg cells from which the nuclei have been removed and observe how the newly formed cells develop normally. However, many people's experiments have failed. But Gordon wasn't convinced, and he didn't have much to do and didn't want to do any more experiments, so he improved on the experiments that others had already done. He asked himself: if the nuclei of adult cells do not work in the late stages of development, can we choose the nuclei of cells that are close in development to embryonic stem cells? If the latter is possible, then we can gradually move on to more difficult adult cells. We have to take our food one bite at a time, and we have to take the road one step at a time. It is based on these ideas. Starting from the simplest point, Gordon took the nucleus of a frog at an earlier stage of development and sucked it through a homemade pipette into a de-nucleated egg. With this unintentional improvement, the experiment worked. Gordon's experiment was quickly validated by his colleagues around the world, rewriting the understanding of development and cellular totipotency. Since then, as the technology has matured and conditions have improved, Gordon has also been able to successfully transplant nuclei from mature intestinal cell sources into frogs.

In the scientific world of the first half of the twentieth century, research abroad was in full swing, while China was in dire straits with wars raging and people suffering. Nevertheless, there was no shortage of Chinese in the international scientific arena. At that time, Tong Dizhou was in Europe, specifically at the University of Brussels in Belgium, following his mentor to learn about nucleus transplantation experiments. Under the microscope, day after day, he would peel off the membrane from outside the frog eggs and try to transplant the nucleus of the cell, becoming skilled to the point of pure perfection. In contrast to Gordon's situation, Tong was considered by his teachers to be a genius, possessing the necessary talents for someone engaged in biological experimentation. On his return to China, Tong, with the support of his wife Ye

Yufen, continued his experiments on amphibian nuclear transplantation, which were at the forefront of international research. But at that time, the domestic scientific research conditions were extremely difficult. A microscope is essential equipment for carrying out experiments. He was in the thrift market to find second-hand equipment, but the price was too high. Even after bargaining with the boss, he would need two years of his salary to come up with the money. Finally, he found a relative who could lend him the money, allowing him to purchase the microscope. Without a stable light source, using a microscope is not helpful. He could only use sunlight to make observations, and on days when there was no sunlight, he used either a kerosene lamp or the reflection of snow on a snowy day. The bright light reflected from the white snow was used as a light source. In this way, despite the almost impossible circumstance, starting from the domestic conditions, he realized nuclear transplantation of fish for the first time in the international arena by using the cells of amphioxus as the research object. As the pioneer of cloning technology in China, Tong Dizhou is also known as the "father of cloning in China." His wife and he are honored as the Curie couple of China.

lancelet

Tong Dizhou

Tong Dizhou and Gordon's situation is relatively similar, that is, their study in childhood. On May 28, 1902, in Tong Jiazhao Village, East Township Yin County, Zhejiang Province, Tong was born into a peasant family. As a child, he studied with his father in a private school, whereas still at the age of young, his father died. He then was brought up by his older brother. At the age of 16,

Tong Dizhou enrolled in Ningbo Normal School and, a year later, transferred to Ningbo Xiaoshi Junior High School. However, in his first year, he scored at the bottom of his class, so he secretly decided to fight for his honor. The next year, he raised his grades to the top of the list and has been the child of others in the mouths of other people's parents ever since. His grades have always been among the best. A year after graduating from high school, he took the entrance exams for Peking University and Southeast University but unfortunately failed. He had to be a special auditor at Fudan University, where he was finally admitted to the Department of Philosophy at the age of 21. However, by chance, after listening to Guo Renyuan's lectures on cat and mouse experiments, he became interested in science and often attended physiology classes. He has since followed the path of biological research.

The theoretical idea behind the scientific research of nuclear transplantation is simple, but the practical applications are far-reaching. At the Roslin Institute near Edinburgh, about 500 kilometers from Cambridge, England, there is a researcher named Ian Wilmut who has pioneered the cloning of mammals. This was done 30 years after Gordon's first report of successful amphibian cloning. The two men started from very different places. The former was as a result of stargazing, while the latter was as a last resort. Wilmut was born on July 7, 1944, in Hampton Lucy, Warwick, England, and grew up at a grammar school. Throughout his time at university, he had been an ordinary student, dreaming of becoming a farmer. However, he had obtained a PhD from University of Cambridge at the age of 27. When he first came to Roslin Hills after graduating, he decided to set up his own laboratory here because he liked the life here. The winters were cold and long, but he could drink whiskey in peace and quiet. At night, he could look up at the starry sky and count the stars. These good times did not last long, however, and soon came to an end when the British government drastically cut research funding and the institute's support fund shrank by two-thirds. He and the other staff at the institute finally realized that research was not just about stargazing, but had to be translated into real technological achievements and money. They understood that it was only when there was more funding that they could talk about their ideals.

It was at this time that he was given a research project by the institute to obtain a transgenic sheep using genetic manipulation techniques that would allow the sheep to produce a special protein that could be used as a drug to treat lung cysts. After months of work, he finally obtained such a sheep and named it Tracy. By collecting blood from this sheep, it was possible to purify and isolate the special medicinal protein. However, it was clear that one sheep was not enough for large-scale industrial production. So, the institute and the company that funded the research hoped that Wilmut would be able to produce more. But this was easier said than done. He had already genetically modified 1,000 sheep embryos for this one, Tracy. He only succeeded with this one, and only by chance. If he wanted to expand production, it would be difficult with just a few sheep, let alone hundreds of dozens. At that time, international research on embryonic stem cells had begun to surface, especially the successful isolation and establishment of mouse embryonic stem cells. This made him think of isolating the embryonic stem cells of sheep. If it could be successful, as long as the expansion of Tracy's embryonic stem cells, more Tracy could be obtained. However, such an endeavor proved to be extremely difficult. Although he was able to isolate the sheep embryonic stem cells, he was simply unable to maintain them in vitro. Soon, these cells differentiated uncontrollably into other cells, losing their original properties. In fact, to this day, no one has been able to establish embryonic stem cells in sheep. He wryly attributes this to the fact that sheep are one of the most tender animals. As a common saying among English shepherds goes, sheep spend most of their lives looking for new ways to die, so of course their cells are no better.

A turning point came in the winter of 1987 when he attended a conference of the International Society for Embryo Transplantation. At the annual meeting, he learned that a Danish colleague had succeeded in taking cells from a cow embryo at an early stage of development. However, these were not embryonic stem cells, and injecting them into an egg to create an embryo that continued to develop for many days was not successful. The news shook him to the core, and memories of Gordon's work 20 years earlier played like a film in his mind. Based on their success, the selection of early-developing cells for nuclear transfer experiments had the potential to be as successful in cloning mammals as it had been in amphibians. Because of the successive failures of his American colleagues in cloning cattle at the time, the directors of the institute did not look favorably on his research program and thought he was far behind in the field. That's when he met a key partner, Keith Campbell. Campbell, an unkempt and thoughtful developmental biologist with long hair, had spent his early years traveling to places like Yemen and Sussex, studying clawed toads. During a discussion between the two, Campbell came up with the novel idea, based on his own research experience, that to achieve successful nuclear transplantation, it would be desirable for the cell providing the nucleus and the stripped egg to be in the same cell cycle. The cell cycle is the process by which the genetic material inside the cell, which is like a rope, periodically loosens and tightens as the cell divides into two. When two cells are in the same phase of the cell cycle, like two people dancing a ballet, perfect harmony can only be achieved if they are in tune with each other. Following his suggestion, Wilmut first tried the nuclear transfer of sheep cells at the undifferentiated stage. After testing 244 embryos, he succeeded in obtaining two Welsh goat lambs named Megan and Morag. He then led a team using the same protocol to experiment with more developmentally mature sheep mammary cells. This time, out of 277 embryos, there were 29 selected as nuclear transfer cells and implanted into surrogate ewes. One of the successful births was the aforementioned Dolly. This monumental day was July 5, 1996, almost 30 years after the birth of Gordon's cloned frog.

If there are people born with a silver spoon in their mouth, then Dolly was the sheep version of them, an aristocrat in the sheep world and, indeed in the animal world as a whole. She had lived in the media spotlight since the day of her birth and attracted public attention with every move she made. The journal *Nature* contacted Wilmut's team early on, hoping to publish their findings in the journal. However, even before the paper was officially published, the news had spread. People from the media, TV stations, newspapers, and magazines drove up in their cars and surrounded the farm and institute with their lonlens SLR cameras. From then on, Dolly went from England to Europe and around the world, becoming a household name. Although the general public didn't understand the significance of it all, with the scientists' confirmation and the global media's spin, people couldn't help but follow the trend. After several seasons, Dolly didn't live for ten years or more like a normal sheep, but she did suffer a lot, including cancer and severe arthritis. She died on Valentine's Day 2003, six years after she was born. Since ancient times, all the beautiful things have had an unfortunate fate. It seems that beautiful Dolly was no exception. As the first cloned mammal in the history of humankind, her significance is undoubtedly huge. Since then, there have been cloned cows, cloned horses, cloned dogs, and so on. Companies have already begun to provide cloned pets for those with the financial means and sentimental attachment to the service. Two years ago, the Institute of Neuroscience of the Chinese Academy of Sciences used highly efficient nuclear transplantation technology to produce cloned monkeys named Zhongzhong and Huahua.

Ian Wilmut   *Keith Campbell*

Among the creators of Dolly, it is clear that Wilmut was in the capacity of an organizer, while Campbell was responsible for the entire framework of biological theory. Since the cloning of mammals, they have been obsessed with the title of the person who first attributed the technology, which later put them in endless quarrels and discord. A perfect research team eventually fell apart, making people sigh pitifully. It is sad that this achievement, which should have won the Nobel Prize in Physiology or Medicine, has been shelved due to the ongoing controversy.

Having cloned animal cells, the next step is to clone human cells. One of the first people to attempt this was Hwang Woo Suk of South Korea, but aspirations are aspirations, and cloning human cells by nuclear transfer is a pipe dream. Hwang Woo Suk was born in 1952 in Chungcheongnam-do, South Korea, a three-minute drive from Seoul. When he was a five-year-old child, his father died, leaving his mother to raise five children alone. The family's poverty made it difficult for them to have enough to eat. For such a family, cattle were an important member. In ancient China, cows were also important domestic animals. As important farm laborers, they could not be slaughtered at will or the killer would be punished by the government. As the family herdsman, the essential daily feeding and cleaning of the cattle pens created a deep bond between the young man and the cattle. It was his dream to become a veterinarian when he was young. With his dream in mind, he was admitted to Seoul National University College of Medicine and became a student at the College of Veterinary Medicine. As with the college entrance exam in China, it was not easy for him to get into the best university in Korea amid Korea's highly competitive college entrance exams. After completing his PhD in animal reproduction in Korea, he spent a short time at Hokkaido University in Japan, where he was introduced to embryos and their development. From there, he entered the field of cloning. On his return to Korea at the age of 41, he succeeded in obtaining the first test-tube cow produced by in vitro fertilization. Two years later, he succeeded in dividing a fertilized egg of a cow into two parts and producing twins. In another two years, he succeeded in obtaining the first cloned cow in Korea by copying Wilmut's nuclear transplantation technique, which was widely reported in the media.

A series of successes gave Hwang Woo Suk the confidence to start experimenting with human cell cloning. In addition, long hours at the front of the laboratory enabled him to skillfully use clamps to hold egg cells. He then used glass pipettes to remove the nuclei from them before injecting other cell nuclei. As a result of his technical skills, he not only became an idol to the other staff in the laboratory but also, after his rise to fame, a figure of great interest to the public. He usually modestly attributed this to the Korean habit of using metal chopsticks. In 2004, *Science* magazine reported that he and his team had succeeded in establishing the world's first cell line for human cell cloning. This made him not only the center of attention in the scientific community but also a hero in South Korea. The president personally presented him with the country's highest honorary award. He was given special bodyguards, and wherever he went in Korea, he was no longer the veterinary surgeon of his early years, but a full-fledged star. Fans asked for autographs and photos even when he was eating in a restaurant. Not only that, but he received unlimited financial support from the government and was able to carry out any research project he wished, something unimaginable in any other country. And in order to establish the image of a scientific and technological powerhouse in the international arena, the government really needed to set a positive example. He happened to come along at just the right time, becoming for a while a symbol of patriotism and national pride for the general public. But the plot was soon reversed. According to the procedure of nuclear transplantation and the calculation of the probability of success, even a 1% success rate is already a very high probability for this technology, especially for the cloning of human cells. Therefore, in order to obtain a successful cloned human cell, hundreds of human eggs will also be needed. This means that hundreds or thousands of women will have to be involved. It is not a small number, especially at a time when there are restrictions on human embryonic stem cell research in the United States. Ongoing investigations into the source of human eggs and continued reporting by team insiders and collaborators gradually brought the truth of the matter to the surface. Not only was the procurement of eggs unethical, but there was no such thing as the creation of human cloned cells. The cells he was presenting to others were human embryonic stem cells derived from the development of normal fertilization of sperm and egg under in vitro conditions. It often takes years or even decades to construct a building, but it only takes an instant to collapse it. Lies and deception cannot hide the truth. With the president's suicide, all the honor of the ruling world from the scientific and political community suddenly faded. Hwang Woo Suk stepped down from the altar, returning from hero to commoner.

Today, cell cloning technology is relatively mature, but it doesn't cover all animals. Although theoretically there is no difficulty, research and application are mostly limited to economically valuable animals due to the bundle of economic interests. Nevertheless, the cloning that people are most interested in and concerned about is still the cloning of themselves. For ethical reasons, there has been no publicly reported human cloning in scientific or political circles. As technology has become more sophisticated, cloning is no longer limited to the development of nuclear transplantation technology. The creation of iPS has made cloning accessible to the general public. With the availability of necessary materials and sufficient funding, the question whether or not human cloning has been carried out by illegitimate groups and organizations requires the attention and scrutiny of society as a whole.

Sperm Meets with Egg

*T*he legend of the meeting of Cowherd and Weaving Maiden at the Magpie Bridge is part of Chinese folklore and tells the story of the legendary Weaving Maiden, the daughter of the Celestial Emperor. She was very good at weaving but only knew how to work hard and not how to dress herself. So, her father took pity on her and allowed her to wander the earth. She then met Cowherd by the river. After they married, Weaving Maiden indulged in her personal life and neglected her work of weaving. This angered the Celestial Emperor, who took her back to Heaven and only allowed her to meet Cowherd once a year on the seventh day of the seventh month of the lunar calendar. The bridge for their meeting was built with the bodies of flying magpies. This is why the seventh day of the seventh month of the Chinese lunar calendar is also known as Oriental Valentine's Day in China. This legend also shows us that love, which is so common for ordinary people, is unattainable for some.

Among the many sciences that have challenged religious doctrines throughout human history, the first scientific technique in the cellular field was in vitro fertilization. As its name suggests, in vitro fertilization is the artificial

fertilization of a sperm and an egg in a laboratory flask. It is then transferred to the woman, which begins to develop normally. Since the invention of this technique is mainly intended for patients who are unable to have children by traditional means of reproduction, it has been fiercely opposed from the very beginning. The main idea behind the opposition is that it goes against the laws of nature and, even worse, that the babies born from this technique are compared to the offspring of Pandora. This shows the worries and fears people had about the technology. However, after almost half a century of practice and standardization, people have fully accepted the use of this technology in medicine, bringing hope and laughter to countless families. When it comes to sperm and egg cells, it can be said that these two cells are the most important of all cells in terms of function and form. They are cells with very different morphologies. In the case of the egg, it is formed in the mother's ovary and when it matures, it is released from the ovary into the fallopian tube. The sperm is an extremely strange cell in shape, it has a big head and a long tail. The most familiar analogy is the tadpole, but the tadpole's tail swings from side to side, pushing the tadpole forward. Whereas the sperm's tail is a propeller, generally rotating, pushing the sperm forward rapidly. We know the story of the tadpole looking for its mother from Chinese primary school textbooks. While the sperm, once it has left the father's body for the mother's, starts looking for the egg without stopping for a moment. The interesting thing is that there is usually only one egg at a time, while there are hundreds of millions of sperm cells. In order to win the favor of the egg, all the sperm have to rush forward as fast as they can. Once the first sperm finds the egg and enters it, the other sperm cells can only "sigh with regret" and die helplessly. This is the story of fertilization in the body, and it is through this fertilization that the human race is able to reproduce and flourish.

The success of in vitro fertilization was not so simple, however, and it was largely due to the lifelong efforts of Robert Geoffrey Edwards. Edwards was born on September 27, 1925, in a small mill town in Batley, Yorkshire, England. His mother was a simple machinist in a mill in the town with a belief that ones can only be successful if they are serious about their studies enough. The strict upbringing of her three children and her passion for nature that she would take them to the neighboring farms whenever she had the chance, contributed to the love of nature and indomitable character Edwards developed from an early age. This served him well throughout his life. At 18, he would have been old enough to go to university, but the outbreak of the Second World War forced him to do his military service for four years. It was only after he was discharged that he went to the University of Wales to study for an undergraduate degree in agronomy. However, within a few hours, he was bored and wanted to switch to the zoology department. But at 26, he didn't have enough money to stay in school forever, and the new department wouldn't allow him to take enough credits to graduate on time. So he transferred to the University of Edinburgh, where he studied and worked part-time. There, he met his true love, Ruth Fowler, whom he married and with whom he had five daughters. Fowler's distinguished family background surrounded Edwards with dignitaries with various titles and famous scientists in various fields, such as the Nobel Prize-winning chemist who complete the isolation of isotopes and radioactive substances, which both honored and pressured him.

Edinburgh was a blessing in disguise for Edwards. After graduating, Waddington's esteem and scholarship support enabled him to complete a three-year PhD studying early mouse development with Alan Beatty. It was at this point that he was introduced to eggs, sperm, and embryos, to which he has devoted his life. The research and knowledge gained during his undergraduate studies laid the foundations for his later work, two of which were crucial in overturning the understanding that adult females are unable to ovulate excessively and that hormones are essential for egg development and maturation. During his six years in Edinburgh, he had not only been very successful, publishing up to 30 scientific papers but had also begun to venture into the field of ethics. Had he continued in this direction, the history-changing technology of in vitro fertilization might have come sooner. However, fate never seems to work out the way it is supposed to. It plays tricks on people.

Sometime later, Edwards moved to the California Institute of Technology, where, under the supervision of Albert Tyler, he began to study sperm-egg interactions and the immunological mechanisms involved. Their research, largely funded by grants from the Ford and Rockefeller Foundations, was not aimed at improving fertility but at finding a way to control the fertilization process from an immunological point of view in order to achieve contraception. For four years after his return, he continued his research in this direction at the National Institute of Medicine. It was still quite fruitful, but he was still nostalgic for his earlier research on the egg and fertilization. So he spent his days researching contraception and his nights researching fertilization.

By this time, his international colleagues had made rapid advances in the field of fertilization and development. He realized that his earlier work on fertilization in mice could potentially be applied to humans, and so his research began to shift entirely toward fertilization. In particular, when he discovered that the eggs of mice, rats, and hamsters all mature spontaneously in vitro, just as they do in vivo, he thought that if the same was true of human eggs, then it would be possible to achieve artificial insemination in vitro. However, for a person involved in basic scientific research, it was never easy to obtain human eggs. Fortunately, on the recommendation of a senior professor at the institute, he was able to meet Dr. Molly Rose, an obstetrician and gynecologist at Edgware General Hospital, not far from the institute. With her help, Edwards was able to obtain enough eggs to carry out his experiments on and off for the next ten years. However, the good times were short-lived, and before he had fully completed his human experiments, the news reached the ears of the director of the institute. With some fanning the flames, he was given a serious warning that he was forbidden to carry out human in vitro fertilization research at the institute.

He had to leave his original institute and accept John Paul's invitation to come to the University of Glasgow to study rabbit embryonic development. It is said that he succeeded in isolating rabbit embryonic stem cells long before the report of Evans on the isolation of mouse embryonic stem cells. A year later, he returned to University of Cambridge to continue his research. To speed up his research, he collected eggs from almost every animal in the neighborhood, including pigs, dogs, sheep, monkeys, and so on, and confirmed his previous findings that eggs matured spontaneously in vitro, but that the larger the animal, the slower the eggs matured. Problems arose in continuing his research on humans: first, he had no stable source of human eggs; and second, the sperm in the semen was not capable of fertilization and needed to be energized in a woman's body in order to be capable of fertilization. To solve the first problem, he contacted a friend he had met on an errand in the United States and indirectly obtained some eggs, barely enough. He also cast a wide net throughout the United Kingdom, seeking and receiving help from several surgeons. To solve the second problem, he met his most important life partner, Patrick Steptoe from Oldham General Hospital in Manchester. Steptoe is an obstetrician and gynecologist who specializes in laparoscopic surgery, which allows him to retrieve energized sperm from a woman's body in a virtually non-invasive way. As a result of these efforts, he was able to achieve and report the first in vitro fertilization of a human egg and sperm in 1969.

Normally, with better research results, it would be logical to apply for funding to carry out the next step in the research. But when the pair applied for a grant from the UK's Medical Research Council, they were unexpectedly turned down. The reason for this was that their research into fertilization, particularly human in vitro fertilization, was mainly solving the problem of infertile families. At the time, this was not only not a clinical-scientific disease worth solving but also exacerbated the problem of overpopulation. In addition to this, research on human beings, after the baptism of the Second World War, was naturally associated with the medical human experiments of the Nazis in Germany. So at that time there was a kind of spontaneous rejection of human cell experiments of the underground rules, the public even compared it to human guinea pig experiments. For the next several years, Edwards had to commute between Cambridge and Manchester for his scientific research. The distance between the two places was more than 200 kilometers, which was a long way to travel with the transport available at the time. On the other hand, he spends a lot of time explaining and debating with the public, the media, and his colleagues.

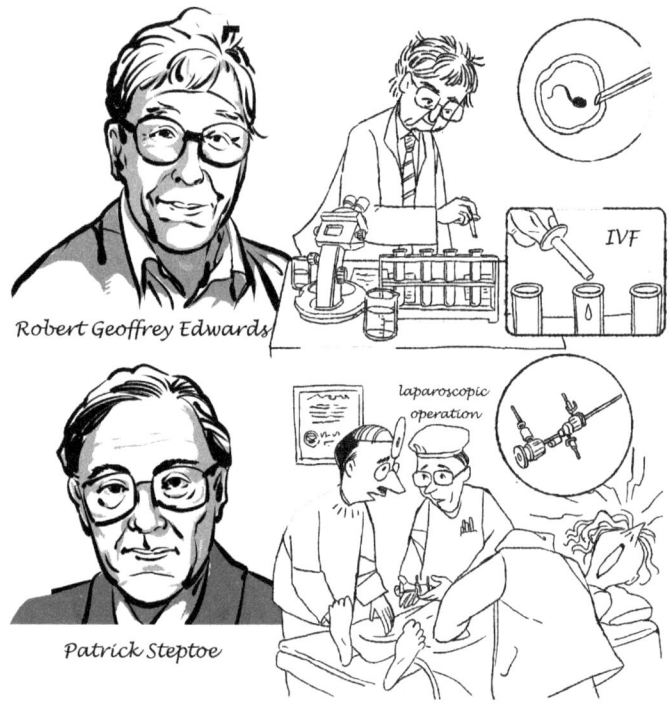

Robert Geoffrey Edwards

IVF

laparoscopic operation

Patrick Steptoe

Thanks to their tireless efforts, the first test-tube baby in human history was born through in vitro fertilization on July 25, 1978. In fact, both in vitro fertilization and the early development of fertilized eggs in vitro take place in Petri dishes rather than in test tubes. The former is more commonly used in cell biology experiments, and the latter in chemistry experiments. However, it is only the media's misinformation campaigns that have led to the popularization of the term "test-tube baby." After years of observation, in vitro fertilization babies (IVF babies) are no different from normal babies, except that the success rate of IVF is not yet 100%. Nevertheless, millions of IVF babies have been born worldwide. As the father of IVF, or more technically the father of assisted reproduction, Edwards was finally recognized with the Nobel Prize in Physiology or Medicine in 2010. Unfortunately, Steptoe died in 1987, otherwise the medal would have been half his.

However, as mentioned at the beginning of this chapter, instead of bringing Edwards any glory, the advent of in vitro fertilization and the birth of IVF babies was initially met with all sorts of repression and booing from both the religious community and the general public. The birth and maturation of in vitro fertilization have, in turn, further spawned the development of other technologies in the field of reproduction. Widely known terms such as sperm donation, egg freezing, and artificial surrogacy have emerged as a result. So, what do all these terms mean and what's the story behind them?

Adult men produce a large amount of sperm every day. Mature sperm will either overflow or die of their own accord, so there is no harm to the human body if sperm is donated like blood. So, what is the point of donating sperm? The main purpose is to provide sperm to men from families who are unable to properly impregnate their children due to congenital sperm defects or reduced sperm motility produced later in life. Our society, therefore, relies on hospitals to set up sperm banks of various sizes. However, not everyone is eligible to donate sperm. People with congenital hereditary diseases and acquired malignant diseases, as well as those who do not like to exercise and stay in their rooms all day, should not participate in the fun. Of course, all donation information is confidential because sperm donation is different from blood donation. Sperm donation will produce offspring, and from a biological point of view, the donor is the biological father of the child. If there is no protection of privacy, it will inevitably lead to unnecessary social chaos, such as the various upheavals that occurred in the early days of the establishment of sperm banks, including the so-called sperm of Nobel Prize winners, PhDs, and celebrities. In addition to sperm donation, people who have high-risk jobs, such as long-term exposure to radioactive materials, toxic substances, and chemicals, or men who are temporarily unable to have children, can freeze their sperm in case they want to have children in the future.

Although today's sperm banks have emerged as a result of the development of in vitro fertilization technology, they are primarily designed for in vitro fertilization. However, the first attempts to freeze sperm and establish sperm banks date back a century or two, to 1776 when the Italian Lazaro Spallanzani discovered that freezing sperm in snow and ice could keep them active for a short time. This so-called freezing attempt was more of a cellular preservation and applied not only to sperm but to other cells as well. The actual concept of freezing sperm was developed by Montegazza in 1866, primarily as a service to soldiers going off to war. Their widows could use their husbands' pre-frozen sperm if they were killed in action. In the 1930s and 1940s, as cell cryopreservation technology matured, sperm freezing became a reality. The first international report of a successful pregnancy using frozen sperm was in 1953. However, the ensuing ethical controversy and lack of legal regulation meant that sperm freezing remained unregulated until it was recognized in 1963, bringing sperm banks into the public eye. The first human sperm bank in China was established in 1981, led by Professor Lu Guangxiu. The first baby inseminated using frozen sperm was born in 1983.

In contrast to sperm freezing, oocyte cryopreservation is usually used by women who have delayed childbearing in order to provide value-added services of their own. They are generally not patients with diseases and cannot be described as egg donors. Due to the physiological cycle of women, the optimal age for childbearing is in their 20s and 30s. However, as the concept of gender equality takes hold, more and more women are entering the workplace. With the need for various title promotions and career development, they often miss the optimal age for childbearing. The quality of their eggs decreases year by year as they get older. Therefore, in order to enjoy safer and healthier fertility in the future, many women choose to freeze their eggs at a young age, just in case. A simple online search will reveal that many celebrities have frozen their own eggs. Of course, there are also women who have to undergo radiotherapy to the ovaries, removal of the ovaries and fallopian tubes because of tumors or other diseases. They freeze their eggs before treatment, also to prepare for normal fertility in the future.

Freezing sperm and eggs is not the same as putting food directly into a fridge at 4°C or −20°C, as we do in our daily lives. In order to achieve the purpose of long-term freezing and to cause as little damage as possible to the frozen sperm and eggs, it is necessary to add certain protective agents, such as glycerol or dimethyl sulfoxide, to the sperm and egg solution prepared for freezing. The main reason for using these chemicals is to prevent the liquid from forming ice crystals during rapid freezing, which can damage the microstructure of the cells. Immediately afterward, the cells were cooled by gradients of 4°C, −20°C, and −80°C, respectively. They were then placed in liquid nitrogen where the temperature could be reduced to −196°C. After the above operations, the cells can remain there for as many years as they wish. When needed, they can be taken out and placed directly into 37°C warm water, then resuscitated and used. Cells in such a freezing process are the true meaning of a frozen age. Time has no affection for them, whether it is a few years or decades later. The awakening of the cell state and its frozen state are almost exactly the same.

In addition to the freezing of sperm and eggs, the freezing of embryos resulting from in vitro fertilization and in vivo fertilization is now a widely used approach in assisted reproduction. Since the birth of the world's first newborn from a frozen-thawed embryo transfer in 1983, the use of this technique in fertility preservation seems to have gradually overtaken the previous two. But as the technology has matured, frozen embryos have brought with them additional legal and social issues. In 2013, China's first case of a dispute over the right to ownership of frozen embryos took place in Yixing, Jiangsu Province, sparking widespread public debate. In this case, a young couple received several fertilized embryos through assisted reproduction at Gulou Hospital in Nanjing due to fertility problems and froze four of them. However, the day before implantation, the couple died in a car accident, raising the question of ownership of the frozen embryos. Did they belong to Gulou Hospital or to the couple? According to the previously signed informed consent, they should belong to the hospital. However, from the point of view of ethics and affection, they should belong to the couple's relatives. The final decision of the court favored the latter. This issue is only the tip of the iceberg of the frozen embryo phenomenon in China, and there are many other issues that have either not yet surfaced or have not yet been legalized. For example, today's embryo freezing and storage centers have a large number of surplus embryos that have not been used for a long time. Up to more than 50% of them have lost contact with their owners. This leads to a large number of medical resources being consumed by these frozen embryos, and their future fate, whether to be discarded or donated for scientific research, faces a series of problems such as legal challenges and regulation.

The development of assisted reproduction technology, while bringing happiness to countless infertile families, has inevitably raised a number of legal and ethical issues. Fortunately, as discussed above, with the development of a sound system, it is basically possible to eliminate the various negative issues raised by this technology. This will lead to its positive and beneficial development. However, the resulting artificial surrogacy is an extremely sensitive issue, both technically and socially.

Artificial surrogacy mainly refers to the performance of in vitro fertilization. Instead of transplanting the fertilized egg into the uterus of the egg donor, it is transplanted into the body of another woman, who carries the pregnancy to term. Modern artificial surrogacy is more like renting a womb and was first documented in the United States in the 1970s. A couple found a surrogate through an anonymous advertisement and had a child through a pre-contract and artificial insemination. In China, surrogacy is explicitly prohibited as it raises a number of sensitive issues, including legal, ethical, health, and even social ones. For example, surrogacy involves the infertile couple and the surrogate mother, making it hard to determine the biological parents of the child. Some consider the mother who gives birth to be the mother, while others consider the couple who provide the embryo to be the parents.

If surrogacy poses the problem of multiple mothers, the latest technology in assisted reproduction, mitochondrial transplantation, has taken the problem to a new level. Babies are now born with a biological father and two genetic mothers, hence the term "three-parent babies." Why does this happen? The mitochondria, which we introduced in chapter 3, not only function as the engine of the cell but also have the property of containing genetic material. If abnormalities occur in this genetic material, like the genetic material in the nucleus, they can be passed on to the offspring, leading to congenital birth defects. In 2019, a couple from Greece were very keen to have a child of their own. However, the quality of the woman's eggs was so poor that both natural conception and in vitro fertilization ended in failure. The solution was to transplant the nucleus of the egg into an enucleated egg provided by another woman, fertilize the new egg with the father's sperm, and implant the fertilized egg into the mother's womb, resulting in the birth of a baby. This baby has genetic material from both parents and mitochondria from the egg donor in every cell, making it a true three-parent baby with two mothers and one father. Strange as it may seem, he's not alone. Before him, in Mexico in 2016, a mother with a mitochondrial genetic disorder used a similar procedure to give birth to a healthy baby boy. Of course, the birth of any new technology, especially for humans, is always accompanied by controversy. The birth of three-parent babies is no exception. But we believe that as long as we follow the philosophy of benefiting others, we will be able to stay on a right and historically tested path, even if it is a short period of darkness.

A Medical Revolution in Cell Therapy

$\mathcal{W}$hen it comes to cell therapy, the information that most people learn about it should come from the media, especially in the age of self-media proliferation. In the overwhelming publicity, the word most often used in its content is stem cells. In fact, the cells that can be used for treatment are by no means limited to stem cells, but also include blood cells, immune cells, and so on. To help you understand what cell therapy is all about, in this section we'll take you through the history and latest advances in cell therapy, as well as what cell therapy really is and what's just a bunch of hype.

In fact, attempts to heal by transplanting cells began centuries ago. Of these, the blood system was the easiest to start with. The earliest idea and technique was to transfuse blood from other animals when there was excessive blood loss. However, given the scientific understanding and development at that time, people did not recognize the immune rejection of cells, nor did they understand issues such as infection. This led to many laughable historical tragedies. Nevertheless, these were wonderful attempts at cell therapy, even if they did not have their wish fulfilled at first.

In the last century, especially with the two world wars, the need to treat many injured people has accelerated the development of medicine to some extent. Among these, blood transfusion has benefited from the high demand during this period, gradually improving and maturing. From the early days of whole blood transfusion to the clarification of the cellular components of blood and their functions to the implementation of component transfusion, such as the targeted transfusion of red blood cells, platelets, or plasma achieves a more precise therapeutic purpose. This not only has better therapeutic effects but also greatly reduces the waste of blood. However, the clinical demand for blood still exceeds the supply, regardless of the type of blood product.

When we talk about blood transfusion, we have to mention bone marrow transplantation, which technically should be called hematopoietic stem cell transplantation. This is because the most important cell in bone marrow transplantation is the hematopoietic stem cell, and the most important cell component in cord blood transplantation, which is gradually replacing bone marrow transplantation, is also the hematopoietic stem cell. Therefore, if there is one stem cell therapy that is first in line today, it is hematopoietic stem cells. It is not only the first stem cell to be used in therapy, but also the most mature source cell for stem cell therapy. In addition, hematopoietic stem cell transplantation not only has remarkable therapeutic effects on malignant diseases of the blood system, such as leukemia, and non-malignant diseases, such as sickle cell anemia, but also shows promising results in the treatment of hereditary and non-hereditary diseases of other systems. The field of application has been continually expanded, which is probably what the old tree blossomed for.

In addition to the blood cells that can be used for treatment, the most popular cells in the last two years have been immune cells, especially for the treatment of tumor cells. These are artificially modified T-cells called chimeric T-cells. The advent of T-cell immunotherapy has led to a fundamental reversal of many previously incurable cancers. Not only has it prolonged patient survival by months or years, but it has also achieved previously unimaginable complete cures. In fact, before the advent of this technology, since the 1950s, people have recognized the role of immune cells in the treatment of cancer. However, the technical means at that time were mainly focused on extracting the body's immune cells from the body and culturing them in vitro. On one hand, the cells were expanded, and on the other hand, they were stimulated to wake up from their dormant state. These cells were then injected into the patient's body to achieve the therapeutic purpose. Theoretically, this is a very feasible set of therapeutic strategies, but it lacks effective precision and the killing power against tumor cells is not strong enough. In addition, due to the lack of regulation, immune cell therapy has caused a number of medical accidents. The Wei Zexi incident, in particular, served as a trigger and directly led to domestic immune cell therapy being silenced for many years in China.

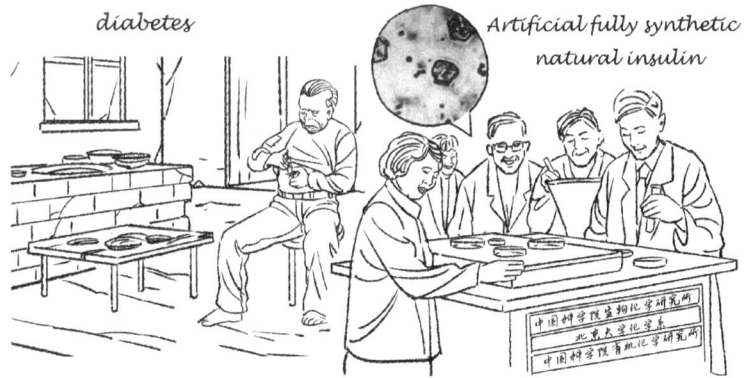

Diabetes mellitus is a chronic disease affecting more than 300 million people worldwide. With appropriate intervention, it is not fatal, but it has become a public health problem due to its serious impact on daily life. One of the main pathogenic mechanisms of diabetes lies in the reduction of beta-cells in the pancreas, which in turn leads to an increase in blood glucose levels due to a reduction in insulin secretion. This is why insulin supplementation is the mainstay of treatment today. In this context, it is important to mention that on September 17, 1965, China synthesized bovine insulin by artificial methods for the first time in the world. This achievement is known as China's earliest and closest to the Nobel Prize in Physiology or Medicine level of natural science achievements. Unfortunately, due to the large number of people involved in this work and the principle of awarding the Nobel Prize in Physiology or Medicine, which states that each achievement cannot have more than three recipients, synthetic insulin's plan to the Nobel Prize in Physiology or Medicine was aborted. Still, this is an achievement of which the nation should be proud, especially in those days of scarcity of food and clothing. It is very admirable that the older generation of scientists were still able to produce world-class research results. However, the drawback of direct insulin delivery is that it requires constant monitoring of blood glucose levels and repeated injections. Therefore, the search for better solutions and treatments is a goal that not only endocrine specialists but also researchers in the field of regenerative medicine are pursuing diligently.

If insulin-secreting islet beta-cells can be replenished at the source, a radical cure can be achieved. Among them, human islet cell transplantation is a good choice. However, it faces the complex problem of post-transplant immune rejection. The most important issue is still the shortage of human islet cell sources. Often, the amount of cells from three donors is only enough for one patient to be transplanted. Therefore, the use of porcine islet cells has become an alternative. In addition, islet cell transplantation using islet beta-cells derived from other cell types is expected to provide a complete solution for diabetic patients in the future. One of the most important factors in the development of the embryonic stem cell field was the hope that many scientists, including Melton, had for a cure for diabetes. However, given the immaturity of the development process for differentiating embryonic stem cells or later iPS cells into pancreatic islet cells, and the fact that safety is not 100% guaranteed, it will be some time before applications in this direction are realized. If it is difficult to supply islet cells directly from outside the body, the direct conversion of other cells in the pancreas that are unable to secrete insulin, such as islet alpha cells, glandular follicle cells, and ductal epithelial cells, into islet beta-cells under in vivo conditions has theoretically good therapeutic effects. A variety of chemical drugs and gene therapies have already been developed on this basis.

Liver transplantation is almost the only effective additional treatment for patients with end-stage liver disease, which leads to cirrhosis and liver failure. However, the extremely limited donor pool severely limits the availability of liver transplants, leaving many patients with liver disease to wait for death. An extracorporeal liver support system with suction-assisted breathing can maintain liver function in patients for short periods of time. However, the source of the most important component of the device, the hepatocytes, has been a long-standing problem. Since hepatocytes themselves cannot be efficiently expanded under in vitro culture conditions, early solutions to populating the artificial liver device have been to use cells derived from porcine livers or hepatocellular carcinoma cells from liver cancer patients as replacement cells. However, these methods are associated with safety issues and dysfunction. To solve the problem of the source of hepatocytes, Hui Lijian's research group at the Shanghai Institutes for Biological Sciences, Chinese Academy of Sciences, was the first to take the international lead in 2011 in turning mouse skin cells into hepatocytes by using the aforementioned technique of driving cells to roll from one foothill to another. Since then, through continuous improvement and optimization of the research protocol, they have successfully programmed human skin cells into hepatocytes. In collaboration with the clinical team at Gulou Hospital, Nanjing University, they demonstrated that an artificial liver device based on these artificial liver cells could increase the survival rate of pigs with liver failure by 80% in pigs that would otherwise live only three days. Building on the success of the large animal study, they conducted a clinical trial that saved the life of a patient with liver failure and restored her liver function to normal levels after treatment. This is a 60-year-old female patient who has had a history of hepatitis B for more than 40 years. She has long-term symptoms such as yellow urine, skin, and eyes. In the late stage of liver cirrhosis and liver failure, there was no available liver source for transplantation. However, with the use of a cutting-edge cell therapy program, she finally turned around from the gates of hell and came back.

Complex artificial liver devices have achieved some efficacy, but are still a long way from full clinical application. This is why researchers are also developing another, simpler way of implanting liver cells. They use physical or chemical synthetic materials to encapsulate the liver cells, preventing them from spreading but allowing their secretions to circulate freely to achieve therapeutic effects. This cell encapsulation technology, from the theory put forward to now, has been developed for more than half a century. It can effectively block the outside world of immune cells and antibodies, as well as other attacks, ensuring the safety of the cells inside the capsule and allowing them to function normally. Attempts at post-cellular encapsulation therapy have been made only in liver failure, and even more promising results have been achieved in the aforementioned treatment of diabetes.

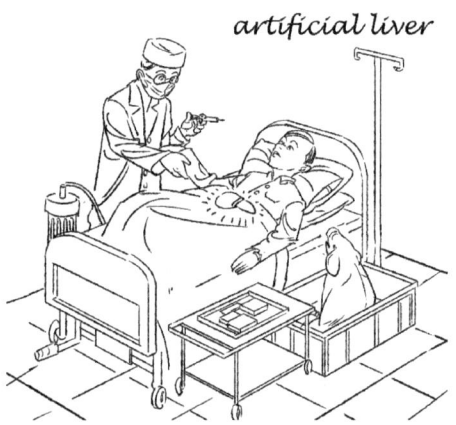

artificial liver

In June 2015, the Burn Center at the Ruhr University Children's Hospital in Bochum, Germany, welcomed a special seven-year-old child to the clinic. This child suffered from Epidermolysis Bullosa, which meant that from the day he was born, he had unexplained blisters of various sizes on all parts of his body, especially on his arms, legs, back, and sides. Six months before he was admitted to the hospital, his condition had deteriorated dramatically due to a bacterial infection. About 60% of the skin on his body had almost fallen off. If left untreated and allowed to progress, it could have led to a serious, life-threatening infection. At this critical moment, after discussion and negotiation with his family, the attending physician decided to use a cellular therapy program that had not yet been formally introduced into clinical practice. This completely innovative treatment option was a kind of compassionate treatment option that can only be offered to patients when there is no other choice, which is currently a recognized procedure of applying new treatment to patients by the international medical community. However, there were still too many unknowns and great risks for the patient and his family. The rescue operation was officially launched after the hospital's ethics committee reviewed the case and the patient's family signed an informed consent form. Three months after the patient's hospitalization, the researchers intercepted a blister from his left groin, where there was no blister. A piece of skin of four square centimeters, from which keratinocytes were obtained by digestion and isolation steps, was cultured and expanded in vitro. After obtaining a sufficient number of cells, two stereoculture protocols were used. One was based on a matrix obtained from vascular isolation and the other was based on a fibrin matrix. These protocols were used to obtain different types of keratinocyte membranes, each with a total area of 0.85 square meters, more than two thousand times the area of the starting skin. During the time between the preparation of the cells and the creation of the artificial skin, the patient loses about 20% of their epidermis. To test the safety and efficacy of the artificial skin step by step, the researchers first performed a trial treatment on the patient's left arm. One month after the artificial skin was applied to cover the area, the skin at the site of the injury not only stabilized without degradation but also regenerated to some extent. Immediately afterward, the researchers transplanted artificial skin onto the patient's legs, neck, right hand, and shoulder. In January 2016, after all the operations, the patient's entire body was covered with artificial skin. There was no more blistering or itching. There is no need to continue applying ointments and other protective treatments. From the time of admission to the time of leaving the hospital, it took only half a year. An almost incurable disease was treated with cell therapy, completely stopping the deterioration of the disease. This is simply a miracle for medical science and a turnaround in life for the patient. At the 21-month follow-up, the artificial skin remained intact on the body, just like normal skin, protecting the child from external infections.

epidermolysis bullosa

skin cell culture

artificial skin grafts

In the case of skin problems, if it is not a large area of damage as mentioned above, but a small area of localized injuries caused by small fights, there is no need to use the methods just mentioned. Otherwise, it will seem like "using anti-aircraft guns to hit mosquitoes—putting a talented person in an insignificant position." For the treatment of common skin problems, we often contact with medicines and daily necessities can be solved, the most common way is to spray, such as mosquito bites after spraying some hydrosol, or sprains after spraying some Yunnan Baiyao and so on. So, if a small piece of skin is missing, can you also spray some cells to speed up the healing process? In fact, some companies have already started to develop such products, using spray cans to spray endothelial cells or epidermal cells directly onto the damaged skin. Though they are still in the research and development stage and animal testing due to low cell viability and other reasons, as a convenient and very attractive treatment, it will gradually mature along with the increase in market demand. Eventually, I believe that one day in the future, cell spray products will be found in roadside drugstores. Do not be surprised at that time.

Blood vessels are one of the most important organs in the human body. They not only extend the function of the heart but also provide a place for blood to settle within the body. With aging and poor dietary habits, blood components such as cholesterol and calcium accumulate in the walls of the arteries, leading to the formation of atherosclerotic plaques. These plaques can jeopardize the function of the organs or tissues in which the blood vessels are located. The most common and deadly consequences are heart attacks and strokes. Treatment for myocardial infarction includes the placement of stents in the heart to widen the blood vessels. Coronary artery bypass graft surgery may also be performed, in which a foreign blood vessel is placed in the diseased area to increase the blood supply. In the early stages of treatment, a piece of vein from another part of the patient's body (such as the leg) was used as a graft. However, this always caused damage to the original area. So, it would be great if synthetic blood vessels could be used as an alternative treatment.

— *artificial blood vessel*

But do not underestimate a small blood vessel, its resilient composition structure is still relatively complex, from the inside to the outside, including several layers, respectively, the inner endothelial cell layer in direct contact with the blood, the middle of the stromal layer and by smooth muscle cells and other components of the outer layer. To mimic these structures, natural or synthetic materials such as silk or chemically synthesized biophilic materials were used in the early days. Techniques such as electrostatic spinning were used to directly wind a hollow tube-like structure similar to a blood vessel. This is a technique very similar to weaving in the textile industry, and Donghua University, famous for its textile major in China, has been very successful in this area. Although these artificial blood vessels have the appearance of nature blood vessels, researchers have found in vivo experiments that they are difficult to use for long periods of time. They lack the necessary toughness and collapse easily. Today, it is possible to mimic natural blood vessels more accurately by using pre-synthesized vascular scaffolds or by isolating blood vessels from animals, decellularizing them to preserve the matrix, and forming a natural vascular scaffold. Human endothelial cells and smooth muscle cells can then be grown in this scaffold. At present, this new type of artificial blood vessel has entered clinical trials for various vascular diseases. It is believed that it will be able to enter the market for the benefit of the public in the near future.

It is the report of one after another exciting success cases that stirs up people's infinite reverie on cell therapy technology. As the third generation of treatment technology other than traditional surgical treatment and drug treatment, there is every reason to believe that cell therapy is in the ascendant and will surely make a big splash in the near future. However, there are two sides to the development of technology, and we must remain cautious in the development of emerging technologies and not make regrettable decisions based on fashion. A few years ago, there were reports that a group of tycoons in China went to Russia together to receive injections of embryonic stem cells in the hope that they could be rejuvenated. It is not known that although the cells can be rejuvenated, it is almost impossible for humans to do so. If rejuvenated embryonic stem cells are injected into the human body, they cannot play a positive role and can lead to serious tumors, which is more than worth the loss.

For the treatment of various diseases, choosing the right cells is only half the work. As a clinical application of cell products, its production requirements are much higher than those mentioned previously in laboratory cell culture. The role of cell quality in cell therapy can be said to be the most important. From the cell plant to the various instruments, materials, culture fluids, and additives used, all of which must meet strict specifications. Otherwise, even if the correct cells are used, they can never be used to treat patients. The construction of the plant and the operation of the cell producers must strictly follow established production standards. The culture of each bottle of cells and the freezing of each tube of cells must be recorded on a daily basis for traceability. National regulatory authorities will also carry out occasional flight inspections of cell production organizations. Materials that come into direct contact with the cells, such as culture bottles, centrifuge tubes, pipettes, must also be from brands certified by industrial and commercial authorities, thus eliminating chemical and physical contamination caused by non-compliant products. For culture media and ancillary components, the most critical point is to eliminate any added substances of animal origin. Fetal bovine serum that is common in the basic cell research should never be used, both to prevent immune rejection and to prevent the spread of bovine spongiform encephalopathy. For those cells that cannot be cultured with serum substitutes, the patient's peripheral blood can be collected and platelet-rich plasma isolated as a nutrient supplement. As long as these culture-expanded cells are fresh, it is best if they can be transplanted therapeutically immediately. However, it is also very delicate if conditions do not allow for temporary cryopreservation. Unlike traditional freezing, where chemical reagents such as dimethyl sulfoxide are added for protection, it is often necessary to map the conditions and use other safer protective reagents to prevent such chemical reagents from causing adverse reactions in humans. In addition, all frozen cells must be tested for contaminants such as endotoxin and mycoplasma before treatment. These contaminants can cause serious adverse reactions in humans after transplantation. When everything is ready, both the cells that have thawed after freezing and the qualified cells that have just been produced need to be transplanted within the first time. They can only be stored in the fridge at 4°C for a few or ten hours at the most. If the time is too long, these cells that are qualified at the moment will become dead cells, not useful at all. And if no special tests are done, it will not be obvious from the appearance at all. Therefore, if you see a person holding a bag or a tube of cells for half a day and claiming it to be cell products for treatment, do not be fooled. The cells have long been inactivated.

There is a very vivid analogy for cell therapy. The human body is like a car. The car is composed of different parts; human beings are composed of different tissues and organs. When the car is broken, we can go to the 4S store to replace parts for repair; when the human body is broken, in the future, you can also use the cells or the tissues or organs formed by cells and scaffolding materials for repair and replacement. These materials can be replaced as commodities and placed on the shelves at any time for people to use, such as a new shelf-type cellular product known as the "human body 4S store." This analogy is too far ahead of its time, but based on the current rate of development of cellular therapy, this will be fully or at least partially possible in the near future. In addition to this form of commodity service in the shopping mall, there is also a kind of farm-type service that is extremely desirable, which is the humanization of animals, especially pigs, so that the cells or tissues and organs in the animal body are suitable for the human body without causing rejection reactions and cross-species transmission of viruses and so on. This is the charm of xenotransplantation. Of course, as a future mode of repair, the automobile industry has been exploring the self-repairing ability of damaged materials, such as scratches on the surface paint of a car or cracks in a steel pipe somewhere. There is no need for human intervention at all. They will recover and heal by themselves. Human tissues or organs themselves will have cell proliferation, cell migration, cell fate transitions, and other physiological activities that also provide the possibility of self-repair of the human body and will be an important mode of cell therapy in the future.

The Marriage of Cell Therapy and Gene Editing

$\mathcal{W}$hile the treatment of acquired diseases is often therapeutic or even cura-tive through cell transplantation alone, the treatment of congenital hereditary diseases is usually not effective using only the cellular therapies described above. Since the underlying cause of inherited diseases lies mainly in the al-teration of the genetic material within cells, effective intervention requires the manipulation of this abnormally altered genetic material. This includes the deletion or correction of the faulty parts and the replacement of the lost parts. The manipulation of genetic material relies on gene-editing technology, which has been compared to God's scissors.

To better understand the editing of genetic material, we first need to recognize genetic material. There is a saying in China that the dragon begets the dragon, the phoenix begets the phoenix, and the son of a mouse natural-ly knows how to make holes, which refers to heredity. Although the power of heredity and its influence on future generations has long been recognized

Gregor Johann Mendel

both domestically and internationally, the first person to delve into the nature of heredity was an eccentric priest. Gregor Johann Mendel, from the Austrian monastery of St. Thosius, planted different types of peas in the monastery garden in the mid-nineteenth century. The pea seeds had smooth or wrinkled skins and were yellow or green in color. Through several generations of crosses between these peas, as well as meticulously recording the number of different types of peas, and then through relatively simple mathematical calculations, he found that the different characteristics of these peas were regular. This led him to discover the law of heredity and speculate on the existence of a specific substance that controls the characteristics, called the hereditary factor. He is therefore known as the "father of modern genetics." However, when Mendel published his discovery in 1859, seven years after Darwin's *On the Origin of Species*, biologists were still focusing on the theory of evolution and did not take an interest in his discovery until its importance was recognized more than half a century later. Following his findings, the search began to determine exactly what kind of substance this genetic factor was. It corresponded to the main components of life: proteins, lipids, sugars, and nucleic acids. In 1909, the Danish scientist Wilhelm Johannsen named the genetic factors "genes," and at the beginning of the twentieth century, Osward Theodore Avery of the Rockefeller Institute in the United States used a delicate experimental design to transform the different substances of two different types of Streptococcus pneumoniae, the smooth and the rough types, into each other. This targeted the genetic factors into genes. The next step for the scientists was to identify the structure of the nucleic acid. It was already known chemically that nucleic acids are made up of four different groups of nucleotides. Nucleic acids are divided into two types: DNA and RNA. Avery's in vitro transformation experiments confirmed that the genetic material was deoxyribonucleic acid, which we often refer to as DNA.

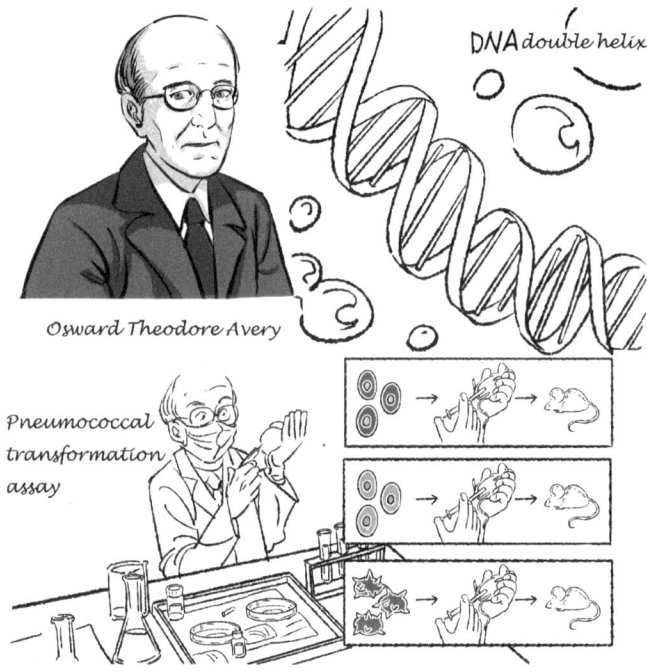

DNA double helix

Osward Theodore Avery

Pneumococcal transformation assay

For substances that cannot be directly observed with the naked eye or a microscope, scientists have to use other tools to detect and then deduce what they look like. The structure of nucleic acids is obtained by analyzing their crystal structure. Crystallographers Maurice Wilkins and Rosalind Franklin obtained the first diffraction patterns of nucleic acids by x-raying the crystals they formed. James Watson and Francis Crick, after seeing the above pattern, first constructed a model of two intertwined nucleic acids with wire and cardboard. This is the famous double helix structure. They reported the model in 1953 and were awarded the Nobel Prize in Physiology or Medicine in 1962.

However, the two scientists who were the first to obtain the important diffraction pattern were denied the prize because of non-scientific factors including gender discrimination. This made it one of the most notorious cases of injustice in the history of science. Half a century later, based on the double helix theory, scientists established the central law and understood how the process between DNA and proteins works. Proteins are made up of more than 20 different amino acids, and these chemicals, called amino acids, have different names. DNA exists to guide the combination of these amino acids according to the sequence of nucleotides already present in the DNA. In order to carry out this process, it is necessary to separate the two chains of nucleic acids that make up the DNA in the nucleus. Then, they must be transformed into nucleic acids that can be transferred to the cytoplasm. Next, the amino acids must be assembled one by one according to the law of three nucleotides that guide the formation of an amino acid. Finally, proteins with a variety of different shapes and functions must be formed. Different proteins control different macro-expressions, such as the characteristics of peas or the smoothness of bacterial surfaces. This indirectly links DNA to these macro-expressions. At the end of the twentieth century, the Human Genome Project completely deciphered the number of nucleotides in human DNA, which is about three billion base pairs long and contains 20,000 to 30,000 genes.

Such a long DNA sequence has to be replicated every time a cell divides. The process of DNA replication is extremely complex and delicate, and with billions of nucleotides, it is inevitable that there will be oversights that lead to changes in the DNA sequence in the cell's offspring. Usually, this change is not fatal to the cell and it spreads from generation to generation as the cell proliferates and divides. In some cases, the change is lethal. The cell itself has a self-repair mechanism that uses its internal enzymes to shear off both ends of the faulty DNA site, removing the faulty sequence and then replacing it with the correct sequence. Of course, any repair mechanism can only guarantee success in the vast majority of cases. With such an enormous number of nucleotides, there will always be some erroneous changes that are neither repaired nor result in the immediate death of the cell. These changes are passed on from generation to generation as the human race reproduces. If these mistakes affect genes that control the production of important proteins, this results in inherited diseases. In the early days, there was both a lack of understanding of such diseases and a lack of basic treatments for them. As a result, there was basically nothing that could be done to treat inherited diseases other than to alleviate the symptoms of the disease. This was not only ineffective but also inefficient. Only human intervention to correct the faulty sequence of genes that cause genetic diseases can provide a cure once and for all.

In order to edit the genes formed by the repeated linkage of four nucleotides, especially those that are genetically incorrectly located, biologists use artificial methods to precisely identify the location of these nucleotides. First, they use the method of shearing to remove the incorrect nucleotide. Then, they use the method of synthesis to add the correct nucleotide. As technology has evolved, there have been a total of three generations of different gene-editing technologies. These technologies identify nucleotides through random positions, then to fuzzy positions, and finally to precise positions. The first two generations of technology both use proteins to identify gene sites. However, one is based on a specific structure in the protein called a zinc finger structure, while the other is based on spaced repeats of specific amino acids that bind directly to different nucleic acid sequences. Using these proteins as guides, biologists have attached a section of enzymes to their tails that can cut genes. This allows them to identify the gene locus they want to edit and then use the enzyme in their tails to cut at that locus, enabling gene editing. The advent of these two technologies has opened the door to gene editing. However, because there are so many combinations of gene sequences, the amount of engineering required to design and synthesize the large number of proteins that act as guides to achieve different edits is immense. At present, all editing technologies have one major weakness or shortcoming—the precision of each editing technology needs to be further improved. Although current technologies can recognize the target, there can be unpredictable interference with bases outside the target. It is completely unknown whether this interference will affect gene function, cell function, or organism function. However, the third generation of editing technology uses nucleic acids to guide the replacement of proteins. This not only greatly reduces the difficulty and cost of the operation, but also allows for the deletion, addition, or substitution of any site on the gene at will. Given the important clinical therapeutic value of this technology, its inventors, Emmanuelle Charpentier and Jennifer Doudna, have also been awarded the 2020 Nobel Prize in Chemistry.

Emmanuelle Charpentier   *Jennifer Doudna*

In 1987, Japanese researchers discovered a set of nucleotide sequences in the genetic material of bacteria that resembled the spaced repeats of amino acids in proteins. They named them CRISPR. In 2003, Spanish microbiologist Francisco Mojica discovered that CRISPR had the same sequence as part of the nucleic acids of viruses in bacteria and ancient cells. He believed this was a new mechanism for bacteria to resist viral invasion. However, the concept was too far ahead of its time, and when he published his discovery, it was often rejected by academic journals. Since then, other researchers have found that

Streptococcus pyogenes absorbs part of the nucleotide sequence of the virus after invasion. This sequence is characterized by intermittent repetition, so that when the same virus invades again, the bacteria can quickly recognize it and degrade the nucleic acid of the virus through the enzyme called Cas. This prevents the virus from replicating and killing it, resulting in a set of bacterial immunity very similar to the immunity of the human body. However, their research was limited to the bacterium's own defense and did not show its potential value and application until Carpentier's intervention.

Born on November 11, 1968, and raised in a small town near Paris, France, Carpentier's dream of pursuing a career in medical research was inspired by an elderly nun she met as a child in an old convent. She obtained her bachelor's and master's degrees from the Université de Paris VI at the ages of 18 and 22, respectively. She then continued her doctoral and postdoctoral studies at the prestigious Institut Pasteur, close to her school. At the Institut Pasteur, a microbiology institute, she began to study bacteria, particularly how DNA fragments play a role in bacterial drug resistance. By the end of her short time as a student, she was happy with the research environment. Although she wanted to be able to carry out her own independent research at the institute, she needed research experience abroad if she was to set up her own independent research group. To this end, she submitted postdoctoral applications to more than 50 research institutions in the US. The advantage of casting a wide net was that she received many replies. From the many offers she received, she was able to find the right one. For the offers, she chose Rockefeller University, where she followed her mentor and began research on drug resistance in Streptococcus pneumoniae. She then moved to the New York University School of Medicine, where she spent two years working with a skin cell biologist, learning how to manipulate genes in mammalian cells. While she learned how some genes control hair growth, she found it much harder to manipulate genes in mouse cells than in bacteria. That's when she realized the importance of developing convenient gene-editing tools.

In 2002, after completing her postdoctoral training in the United States, Carpentier finally had the capital to set up a laboratory in her home country.

Instead of returning to France, she embarked on a wandering research journey, choosing the University of Vienna as her first stop. Over the next seven
years, although she was only able to apply for short-term project funding, her
projects—there are many of them—always managed to get funding to keep
the lab running. Despite the difficulty, the experiments were going well. It was
here that she discovered that one of the nucleic acids in Streptococcus pyogenes had CRISPR properties. To investigate further, she collaborated with
other groups to map the nucleic acid of the whole streptococcus. When the
mapping was complete, she found that the streptococcal CRISPR differed from
the reported studies in that it not only contained the shearing enzyme Cas9
but also involved two unique nucleic acid sequences. This unexpected discovery
quickly made her realize that if she could work out how these three elements
worked together, she could manipulate them, which would be a powerful
gene-editing tool. In 2009, tired of the big city, she moved to the Microbiology
Research Center in Ürmörå, a small, old town in northern Sweden. The center
was young but well-funded, and the long, dark winters made it easier for her
to concentrate on her work. One evening that summer, a student told her that
the experiment had been successful. She suppressed her excitement and told no
one. As a young researcher with no reputation in the field, she needed to keep
working hard and produce more convincing experimental data if she wanted to
get her paper published. A year later, the paper was successfully published, and
it was like dropping a bomb on the gene-editing community. This turned her
from a nobody into an academic star. In 2011, she met Doudna, a structural
biologist at the University of California, Berkeley, at a meeting of the American Society for Microbiology. The two hit it off and began collaborating, soon
reconstructing the CRISPR system in a test tube and realizing DNA cleavage.
This led to the establishment of the CRISPR-Cas9 gene-editing technology,
which was first used in 2013 by Chinese American scientist Feng Zhang to successfully edit DNA sequences in mammalian and human cells. The three were
initially collaborators, but eventually fell out in a dispute over patent ownership. Although Zhang won, he lost the Nobel Prize in Physiology or Medicine.
It was also in 2013 that Carpentier moved her lab to the Hannover Medical

School in Germany, where she was also a department head at the Helmholtz Center for Infection Research. The main advantage of this move was that she was able to hire technicians from her own research group, creating a more stable research team. In 2015, she was introduced to the Max Planck Institute in Berlin, Germany, where she has been working ever since. Along the way, it is the ideals she set at a young age that enables her to persevere in her ascetic research career. She has traveled from country to country, jumped from lab to lab, and spent time at a total of nine different research institutes in five countries over the course of her more than 20-year research career. As her supervisor said of her, "She's so smart, she can set up a lab in the desert."

With a gene-editing tool, the next step is to consider how to get the scissors into the cell and into the body. Here, we have to mention viruses, which everyone is scared of at the mere mention of. As a virus, its natural advantage is that it can easily enter the cells it wants to enter and set up camp. It does this either by directly expressing proteins or by directly inserting the genes it carries into the genetic material of the cells. This way, it is always present and can perform a function when the cells proliferate or divide. Like removing the fangs from a poisonous snake, the snake loses its viciousness and becomes a pet to be kept. In the case of viruses, biologists use the same means to remove their lethal genes but still preserve their ability to invade cells. This allows them to achieve captive breeding of viruses, enabling them to abandon their evil and turn to good for our use. In particular, the use of viruses as vectors for gene-editing tools has become an almost routine way of entering unusual laboratories. Of course, for different cell types, organisms, and gene fragments, the selection and modification of virus types should be tailored to local conditions and not generalized. Among them, the well-known human immunodeficiency virus has been completely tamed after modification and is now used in everyday genetic manipulation. This shows that such a hateful and ferocious virus can also be beneficial to humans.

With the availability of multiple tools and vectors, it is possible to cure a variety of diseases. Let's just say it's too early to say. To date, there are only a handful of gene editing combined with cell therapies in the field, mainly for the treatment of rare diseases. However, the potential of this new biological thera-py has already been demonstrated, and researchers and industry have rolled up their sleeves and are ready to take on the field.

This brings us back to AIDS. Wouldn't it be a bit like fighting fire with fire if we could take the human immunodeficiency virus, genetically edit the cells used for treatment, and then use them to treat AIDS? Although we have not yet fully realized such an idea, two crucial steps have been taken.

The first step is to use cells born with a variant of the CCR5 gene to treat AIDS patients. How is this possible? It has to do with the way human immu-nodeficiency virus infection works. The human immunodeficiency virus likes to grab and burrow into immune cells, especially those that have a protein called CCR5 on their surface. It is as likely to grab them as a bull at a bullfighting festival in Spain who sees a fluttering red piece of cloth and charges headlong. Once inside, the virus kills these immune cells, eventually rendering the body immune deficient. This leaves the infected person vulnerable to a variety of seri-ous infections and malignancies, ultimately leading to death. Nature's ingenuity not only creates species diversity but also diversity within the same genes. Some people are born with a mutation in the CCR5 gene that prevents the human immunodeficiency virus from entering genetically mutated cells. As a result, their immune cells are not afraid of the virus at all. Therefore, if the infected immune cells are removed from the body and replaced with CCR5-mutated cells, the AIDS patient can regain immunity. The Berlin patient, the first AIDS patient to be cured in human history, was permanently cured using this meth-od. However, the probability of finding someone with these mutated cells and the same blood type as the patient is extremely low and depends on luck.

The second step involved gene editing of CCR5 in normal immune cells to achieve the same efficacy as the natural variant, followed by cell transplantation for therapeutic purposes. In this regard, Deng Hongkui and Chen Hu's group from Peking University in China have already achieved positive first clinical results, which are encouraging. In May 2016, they saw a 27-year-old patient with both HIV and leukemia. After one year of cocktail therapy and six cycles of standard chemotherapy, the number of viral particles in his blood dropped to undetectable levels, and his leukemia went into remission. To test the safety and efficacy of gene-edited hematopoietic stem cells in treating this type of disease, they isolated hematopoietic stem cells without the CCR5 mutation from a 33-year-old male donor with the same mating type. They cultured them in vitro for two years, then genetically edited them for CCR5 using CRISPR editing technology. After waiting another two hours, they infused these gene-edited hematopoietic stem cells into a former AIDS patient in July 2017. After 19 months of observation, although the percentage of gene-edited lymphocytes was only about 5%, this patient's leukemia was cured. The cells transplanted could survive for a long time without any adverse effects, which is an initial indication that gene editing is safe. We believe that in the near future, the use of this gene editing plus cell therapy method can completely cure AIDS patients. Cocktail therapy, which once played an important therapeutic role, can be consigned to history.

Anemia is a very common disease, especially in women. Most anemias are due to acquired nutritional deficiencies, such as low iron intake and folic acid deficiency. They can, therefore, be treated by supplementation with elements or factors necessary for hematopoiesis. However, there are several types of congenital anemia—dysglobinopoietic anemia and sickle cell anemia—which are caused mainly by mutations in the genes that make up hemoglobin and globin. This results in the inability of hemoglobin to perform its function properly and triggers erythrocytic hemolytic anemia. There are nearly 100 million carriers of the mutated genes worldwide, and China is mainly concentrated in the southern region, especially in Guangxi, where the carrier rate is the highest at 6.4%. Patients with severe dysglobinopoietic anemia die in utero or within hours of birth. Some require lifelong transfusion or iron-exclusion after birth, and even then, the average survival age does not exceed 10 years.

As this is a relatively simple disease caused by a single gene variant, using gene-editing technology to correct faulty genes in hematopoietic stem cells that can produce red blood cells and then reinfusing the gene-edited hematopoietic stem cells into anemic patients could cure such patients once and for all. Encouraging progress has been made in this direction, both internationally and domestically. In 2020, a team of researchers from East China Normal University, in collaboration with clinical research teams from several hospitals in China, sequentially transplanted genetically modified autologous hematopoietic stem cells that had achieved the reactivation of globin into patients with severe globin-producing anemia. This freed them from long-term dependence on transfusion therapy.

Hemophilia is one of the rare blood disorders that are present from birth and remain with us throughout our lives. It is memorized annually on April 17 as World Hemophilia Day. As early as AD the second century, the Jewish Torah *Talmud* contain an account of a disease in which some male infants who underwent ritual circumcision bled profusely and died. In such patients, even a small cut on the surface of the skin can be fatal when bleeding occurs. Scabbing over a wound, which is perfectly normal for a healthy person, is a fantasy for them. Therefore, if bleeding occurs, such patients are likely to die of blood loss unless immediate action is taken. In addition, hemophiliacs are over 70% more likely to have deformities in their joints, and in severe cases often require replacement of artificial joints. In addition, the high rate of internal bleeding makes surgery for other complications difficult or risky, further increasing mortality. For children with these conditions, the high prevalence of oral disease due to chronic lack of dental care for fear of internal bleeding in the mouth is also a problem that cannot be ignored. In our country, almost three out of every 100,000 people suffer from this disease. While the best treatment for this disease was scalding in the twelfth century, the current effective treatment option is mainly coagulation factor supplementation. This is timely and effective in stopping bleeding, but the high cost of treatment often makes it difficult for most families to maintain it over the long term. On the other hand, prolonged injection of the factor, which leads to the formation of inhibitors, can also make the method increasingly ineffective.

In Europe, the most famous case of hemophilia is probably that of Queen Victoria of Britain. Born in 1819, Alexandrina Victoria became Queen of Britian at the age of 18 and ruled the country for 64 years. During her reign, Britain was extremely prosperous and became known as the empire on which the sun never sets. The period of her reign is also known in history as the Victorian era. Of course, her success could not be separated from the political marriages she initiated through her children. Her nine children intermarried with the nobility of many European countries and gave birth to 35 grandchildren, earning her the title of "Grandmother of Europe." But as she grew politically wealthy, she also spread the then-unknown "blood curse" to the royal families of Europe. Her fourth son died the next day after falling and injuring his knee. Her second daughter, who married a German duke, gave birth to seven children. One of them died of internal hemorrhaging at the age of three from a fall on the balcony. Even her great-grandson, the last Czar of Russia, was pale and weak from chronic hemorrhaging and later bedridden for life from a fall. Quite a few of her other children and grandchildren have been maimed or killed by bleeding after a fall.

The hidden culprit behind hemophilia is a congenital genetic mutation that causes a malfunction in an important protein that controls blood clotting. Hemophilia was first described as an inherited blood clotting disorder in 1823 by the German Scheuren. As an X-linked recessive disorder, it is characterized by being inherited in females and contracted in males. There are three types of hemophilia, depending on the gene that codes for the clotting factor involved: hemophilia A, hemophilia B, and hemophilia C. The first is primarily a muta-

tion in the gene for clotting factor VIII, discovered by the American Kenneth Brinkhous in 1939. The second is a mutation in the gene for clotting factor IX, discovered by American researchers in 1952 in a patient called Stephen Christmas. The disease is also known as Christmas disease. The former is far more common than the latter. Once the core of the problem had been identified, things began to look up. Using gene-editing technology to edit the wrong genes in cells that produce clotting factors, such as hematopoietic stem cells and liver cells, and then infusing these genetically corrected cells back into the body so that the patient produces clotting factors with normal clotting function is potentially the only hope for a cure for hemophilia. There are two main methods for obtaining factors with higher clotting activity: one is to insert the genes of factors that spontaneously mutate in nature but have better clotting functions, and the other is to use artificial methods to modify the encoded genes one by one to find the best combination. It should be noted that overexpression of such factors also carries certain risks and is prone to thrombosis. Therefore, the optimal level of factor expression and the appropriate time of expression have been precisely controlled to maximize the therapeutic benefit. Of course, some people are also trying to adopt the method of virus-mediated expression of such factors by injecting such viruses directly into the body to achieve therapeutic purposes. In 2015, the US company Spark Corporation carried out adeno-associated virus-mediated expression of coagulation factor IX for the treatment of hemophilia B in Phase I and II clinical trials. However, the safety and efficacy of such methods still need to be further verified.

Immune cell therapy, the hottest technology of the moment and the one that has brought hope of life to countless cancer patients, is the proudest and most introduced technology on the board. Technically, this type of therapy also includes cell therapy and gene editing. We already know that immune cells are an important weapon in our body against foreign invasion and self-inflammation. However, tumor cells develop from normal cells in the body, making it difficult for our own immune cells to recognize them. However, there are certain differences between tumor cells and normal cells. After millions of experiments, key differences have finally been identified, such as the fact that leukemia cells have proteins on their cell membranes, more commonly known as CD19, that make immune cells think they are the bad guys. So how do you spot them? Cops can't catch bad guys with their bare hands; they have to be armed with guns, handcuffs, and other tools. For this reason, scientists have increased the equipment for immune cells and are using gene-editing technology to efficiently recognize tumor cells with special proteins. After the transformation of immune cells, the most famous is when chimeric antigen receptor T cells are created. These cells, like a pair of eyes of fire, once in the body, will have the goal to find tumor cells. They achieve high efficiency and precision in killing tumor cells, thus achieving the purpose of treating tumors. This novel immune cell therapy program is known as CAR-T.

The first scientist to develop CAR-T was Zelig Eshhar of Israel. In 1993, he used gene editing to modify T-cells and experimented with their ability to kill tumors. With Eshhar's help, American Steven Rosenberg reported the first case of CAR-T treatment in 2010. The lymphoma patient was not cured, but experienced a significant remission. It wasn't until 2011 that American Carl June reported a case of cure using this method, which set the whole field on fire. June's early research focused on activating and modifying T-cells to treat AIDS patients, but in 1996 his wife was diagnosed with ovarian cancer. She died

five years later, and this left him determined to pursue research into the use of immune cells to treat tumors. However, research requires a lot of financial support, and in the early days, the field was not widely known or recognized, so it was difficult to apply for research grants. Fortunately, a private foundation called the Cancer Gene Therapy Consortium provided $1 million in support. The founders of the consortium decided to set it up and help more patients like themselves after witnessing the death of their daughter-in-law from breast cancer.

As the first person to eat the crab, the American girl Emily Whitehead benefited greatly. From being declared beyond treatment to adopting this therapy, she truly went through a life-and-death ordeal. Aged five, she was diagnosed with acute lymphoblastic leukemia. After two rounds of chemotherapy at the Children's Hospital of the University of Pennsylvania, her legs appeared to have necrosis and she almost faced amputation! Just 16 months later, her leukemia returned. At this point, the only option was a bone marrow transplant. However, at the same time, she had to deal with the huge side effects of the treatment. On balance, her parents decided against this option. The little girl's condition worsened every day. The leukemia cells in her body doubled every day, and she could only undergo another round of intensive chemotherapy. Three weeks had passed and there was still no improvement in her condition. Even her doctor seemed to have given up hope of a cure and suggested that her parents enter hospice care for the last time. This is a spiritual comfort given to patients nearing the end of their lives, allowing them to leave this world with a smile and dignity instead of pain and suffering. But for her parents, it didn't make sense; they wanted to see their daughter live, and if there was a glimmer of hope, they would try. That's when they heard that scientists at the University of Pennsylvania were working on a new type of tumor immune cell therapy, but hadn't conducted any clinical trials. After a lot of hard work, Emily became the first person to participate in this new therapy. But after the third dose of the therapy's healing cells was injected, she developed severe symptoms of high fever, respiratory failure, and shock. As part of the treatment protocol, a sample of her blood was sent for testing to help identify the cause and a targeted relief program. With time ticking away and the situation critical, the attending doctor made countless calls to his colleagues in the laboratory, begging them to speed up the process. Two hours later, the test results came back. The level of interleukin-6 in Emily's blood was 1,000 times higher than normal, so it was no surprise that it was causing such a severe side effect. Finding a way to eliminate interleukin-6 through online searches and inviting specialists to the clinic was the immediate focus and only hope. That's when they remembered that Professor June's daughter was taking an antibody drug for her arthritis, tolizumab, which is an interleukin-6 inhibitor, and that the hospital pharmacy happened to have it in stock. A few hours after Emily was given a shot of tolizumab, there was a marked improvement in her symptoms. She finally woke up from her coma on her seventh birthday, which was the best birthday present ever. At present, Emily is very healthy and can say that she has been cured of her leukemia.

Emily's ability to beat the disease was not only a personal miracle for her, but also a triumph for immune cell therapy against cancer. It should not be

Steven Rosenberg   *Carl June*

forgotten that on September 17, 1999, at the University of Pennsylvania, an 18-year-old boy named Jesse Gelsinger was treated with gene therapy. The field of gene therapy has been in darkness for more than ten years since Gelsinger died after a clinical trial of the same treatment. If Emily had suffered the same tragedy, the fate of immune cell therapy for tumors would have followed suit. This time, luck was on everyone's side, from the patients to the doctors on the front line and the researchers behind the scenes. Since then, CAR-T therapy has blossomed around the world, with cure rates of more than 80% for leukemias for which there was no cure. In 2017, the US Food and Drug Administration approved the clinical and commercial use of CAR-T cell therapy. Pharmaceutical giant Gilead spent more than $10 billion to acquire the technology from Kite Pharma, an early developer of the technology. The problem is that the product costs more than $400,000, an unaffordable amount for the vast majority of patients' families if the cost of treatment could not be reduced.

In addition, it always seems to be the case that success or failure depends solely on a single individual. It is because the gene-edited immune cells exist in the blood system and can run all over the body; therefore, CAR-T therapy for hematologic malignancies has a very good efficacy. However, if this type of immune cell wants to effectively enter solid organs, such as into the liver to kill liver cancer cells, into the brain to kill brain tumor cells, the effects are minimal. Because even if their size is small, it is still not easy to drill into these tissues. At most, one can squeeze the door and probe a head, but entering the whole body is impossible. If we can improve their "bone-shrinking power" in the future, or open the doors to different organs larger, there may be unexpected effects.

Restless Tumor Cells

𝓘f we analyze the problem from the perspective of dialectical materialism, everything can be divided into two with good and bad sides. Just as human beings are divided into good and bad, cells are no exception. We call normal cells "good cells," tumor cells "bad cells," and cancer cells "even worse."

The appearance of tumor cells or cancer cells can be traced back millions of years, and scientists have found fossils of osteosarcoma formation in ancient human fossils from archaeological excavations. In addition, researchers have used modern anatomical techniques to find traces of bowel cancer cells in Egyptian mummies that are still preserved today. This suggests that the person may have died of the disease at the time. Cancer is a curse for the Egyptians, according to ancient Egyptian medical books preserved to this day. As early as 1600 BC, they found a variety of cancers, such as breast cancer. In order to treat them, they tried burning and branding methods, smoke methods, knife-cutting methods, incantations, and other paths. In 400 BC, Hippocrates of ancient Greece observed the manifestations of cancer in superficial places such as the skin and nasal cavity, and found that the cancer cells crawled out from the center and into the normal tissue, looking very much like crabs with claws, so they were named crabs. It was not until the first century BC that the word "crab" was translated as "cancer." In AD the second century, Galen, an ancient Greek, called the lesioned tissues tumor, which would only grow lumps, but with no invasion of normal tissue occurs. In AD the third century, this type of disease was also recorded in ancient China, first in the book *Pulse Classic* compiled by Wang Shuhe of the Western Jin Dynasty (AD 265–317). Chinese medicine calls it "rock," meaning something with an uneven surface and hard texture. This is very graphic and apt.

So, how do tumor cells develop? There are many different theories about the formation of tumor cells. In a nutshell, tumor cells are transformed from normal cells, and the reason for the transformation is either genetic or environmental.

In addition to the genetic characteristics of tumor cells, which are caused by changes in their own DNA code, there is another important external factor: viruses. These are too small to be seen with the naked eye and too small to be observed with a conventional optical microscope. There are millions of different types of viruses, far more than there are cells, both in terms of type and number. Viruses tend to attach themselves to certain types of cells, affecting their properties. The first studies found that infection of normal mammalian cells by the chicken sarcoma cell virus resulted in the direct transformation of these cells into tumor cells that went into a state of unlimited proliferation. Why did this result occur? It was thought that the virus inserted a fragment of its own DNA into the DNA of the normal cell, giving it a different fate, just as the insertion of a plant gives it completely different growth characteristics. We now define this DNA as oncogenes, which, when present and active, cause normal cells to become tumor cells. Based on this idea, in the opposite direction, people have also found genes that can inhibit the infinite growth of tumor cells and called them tumor suppressor genes. Of course, these genes have not been found in viruses but are naturally present in normal cells.

One of the best-known cases is the link between the human papillomavirus (HPV) and cervical cancer. There has been much controversy about pre-

ventive vaccines against tumors, and it is well-established that the HPV vaccine
is effective in preventing cervical cancer.

Harald zur Hausen

Harald zur Hausen was born on March 11, 1936, in Gelsenkirchen, a
coal-mining town in northern Germany, like Datong in Shanxi Province,
China. Due to the Second World War, his schooling was interrupted from
primary school onward. However, he was neither at the top nor at the bottom of his class and survived until the age of 19 when he took his university
entrance exams. As a child, he spent every day with flowers, birds, fish, and
insects, which cultivated his love of nature. So when it came to choosing a subject, he hesitated between biology and medicine. Although he eventually chose
medicine, he attended all sorts of biology lectures and courses during his time
at university. It was hard, but he persisted and gained a lot. After graduating,
he wanted to continue his scientific research; but for financial reasons, he had
to get his medical license first. This led to a two-year internship at a hospital,
where he was introduced to obstetrics and fell in love with the subject. After his
internship, Hausen moved to the Department of Medical Microbiology and
Immunology at the University of Düsseldorf, where he began researching how
viruses cause chromosomal changes and was introduced to microbiological
diagnostics. Seeking professional advancement, he moved to Philadelphia, the
US, three years later to begin his postdoctoral career. There, he started a family,
built a career, and had children. His mentor's research interest at the time was

the newly discovered human herpesvirus type 4, and the entire group was involved in developing assays for the virus as well as epidemiological studies. But Hausen hadn't been trained in molecular biology, and it was difficult for him to keep up with everyone's progress. To make up for his personal shortcomings, he incited his mentor to let him start researching adenoviruses. Although he was extremely reluctant, he did not give him a hard time. At the same time, he did some research on the virus infecting lymphocytes and lymphoma cells, which is considered to be one of the major types of tumors caused by herpesvirus infection. At the age of 33, he finally ended his life abroad and returned with his family to his home country. He then began to build up his own independent research group at the University of Würzburg. This time, he was finally able to conduct research on herpesviruses, which he did for three years.

After achieving a small success in this area, he was offered the position of director of a newly built Clinical Virus Research Institute. In order to make progress, he planned to start a new direction of research. Of course, the new direction had to be based on his previous research background and not stray too far from it. Otherwise, it would be a castle in the air. Since lymphoma was linked to the herpes virus, what about other tumors? In particular, the myths about cervical cancer and its relationship with viruses have been circulating for centuries, especially in the 1960s when the herpes simplex virus type 2 was the most common type of herpes simplex virus in the world. He therefore decided to look in this direction and asked his colleagues to look for traces of the herpes simplex virus in clinical specimens of cervical cancer. However, contrary to expectations, all efforts failed. It was at this point that he became aware of the growing number of reports of perigenital warts turning into cancer. It was already known that the former contained human papilloma virus. So all he had to do was test for the presence of such viruses in cervical cancer. With a lot of help, Hausen eventually found the results he was looking for in several samples. Five years later, Hausen and his team moved to the Institute of Virus Research at the University of Freiburg, where he became the director. Thanks to being in the right place at the right time, he and his team discovered several subtypes of the human papillomavirus and confirmed that the virus causes cervical cancer.

With these results, they naturally thought of using a vaccine to prevent cervical cancer by preventing viral infection. However, after contacting various pharmaceutical companies, they found that none of them were interested. It was not until a few years later that the prevention of cervical cancer through HPV vaccination was accepted. He was awarded the 2008 Nobel Prize in Physiology or Medicine for his original discoveries and contributions in this field.

Having discussed the genetic background of tumors, it is now time to discuss the environmental factors that lead to the production of tumor cells. There is a famous analogy about the relationship between seeds and soil. A seed will not germinate in all soils; it will only take root in the right soil environment. In the same way, a normal cell with a fatal DNA mutation does not always turn into a tumor cell. It is only in a certain inferior environment that it will show its ugly face, which may be the so-called "one who is near the vermilion is red, and one who is near the ink is black." The environment includes both small and large environments. The small environment mainly refers to local inflammatory reactions, such as long-term infections or tissue ulceration. For lung tumors, it is mainly the malignant changes in normal lung cells caused by long-term smoking or inhaling polluted air. The notorious environmental factors are nicotine and dioxins. The change in the general environment is irreversible aging, so we can also consider tumors to be geriatric diseases. Of course, with the gradual aging of the body, the increase in tumorigenesis due to the accumulation of DNA mutations is an inevitable factor. There is increasing evidence that aging causes a change in the body's overall immune environment, which is unable to control the malignant transformation of DNA-mutated cells or the rapid growth of tumor cells that have already undergone malignant transformation. As a result, tumors have tended to overtake non-malignant conditions such as cardiovascular disease and organ failure as the leading cause of death in older people.

The most common disease caused by environmental change is bacterial infection, which causes long-term inflammation and, in severe cases, even tumors.

Barry Marshall was born on September 30, 1951, in Kalgoorlie, Western Australia. It is a gold mining town located about 370 kilometers from the city of Perth. His father was a 19-year-old apprentice locksmith, and his mother was an apprentice nurse who had just turned 18. During his childhood, he moved around with his parents as they changed jobs, which was hard but fun. The first time he moved, the family drove 1,000 miles to the West Coast in a car. However, the car broke down and the family stayed there while it was being repaired. They lived in Carnarvon for four years. As his grandfather and grandmother still lived in Kalgoorlie, they moved back. When he was seven, the family moved to Perth. Over the course of seven years, he had two brothers and a sister and often took them on adventures. Once, he made his brother jump out of a tree, resulting in a broken leg. In primary school, his grades went up and down because he didn't study hard enough. Once or twice, he was first or second in his class, but most of the time, he was in the middle. He spent most of his time following his dad around, tinkering with car engines and fixing cars. He was curious about everything and read a lot of his father's collection of mechanics books and his mother's medical books. And he wasn't just a nerd; he was good at applying what he learned, even if he was often wrong.

The seed and soil theory of tumors

Helicobacter pylori

Barry Marshall

When he was 12 years old, he and his brother were at home taking care of their less than two-year-old sister. They were not paying attention when they found their sister drinking a bottle of milk. However, the contents of the bottle were not milk but kerosene. Not long after, they noticed that their sister was acting abnormally, hiccuping all the time. In a hurry, he called the emergency number while giving her artificial respiration. The artificial respiration turned out to be useless because his sister was still breathing and he could smell the kerosene in her mouth, knowing she had drunk the wrong thing. Fortunately, the ambulance arrived in time to save her life. This incident made him a minor local celebrity for a while. Throughout his youth, his father was the biggest influence on him and several of his siblings. This allowed him to learn a lot between games and try fun things like making his own gunpowder and fixing appliances. After graduating from high school, Marshall entered Newman College and, according to his wishes, he definitely wanted to major in electromechanical engineering. However, there is something to be gained and something to be lost. He used to focus so much on gaming that his math background was too thin, so he had no choice but to choose the medical profession.

At university, he was lucky enough to meet the love of his life, a fellow psychology student, and they were married in their fifth year. Since then, for nearly a decade, his life had been a routine of graduation, internships, and residency rotations. During this time, he and his wife had had four children who took up his leisure time to play with. It wasn't until 1981, when he was on a gastroenterology rotation that he met Robin Warren. The change in his life happened quietly. At the time, Warren had found some curved-shaped bacteria in some of his patients' stomach biopsies but didn't know if they were related to clinical symptoms. So, the head of the department arranged for him to help Warren with his research. He was more than happy to help because he had learned that one of Warren's patients was his neighbor who had been suffering from stomach pains for a long time without being able to find the cause. He had suggested that she see a psychiatrist to see if psychotherapy would help. Thus began a long collaboration in which the two spent countless afternoons together. After observing numerous patient samples and reading a great deal of literature, it was clear that there was a link between bacteria and gastritis. However, they had not been able to determine exactly what phenomenon they were observing. A year later, Marshall received a grant from Fremantle Hospital to set up an independent laboratory and was able to continue his research. Several international research groups began to report the same findings, namely that there was an association between Helicobacter pylori and gastritis. At the same time, there was a lot of skepticism within the academic community, particularly as he was unable to develop animal models to simulate the clinical phenomenon and therefore faced many obstacles in publishing his work.

During those disheartening days, the news from the clinic boosted his confidence, especially that the antibiotic treatment had produced good results. These results also led him to suspect that these bacteria first caused gastric ulcers, then gastritis, and finally further deterioration, possibly leading to gastric cancer. To test his hypothesis, he took some of the isolated bacteria from the laboratory and drank it all in one go, without consulting his wife or seeking the approval of an ethics committee. The effect was immediate. A gastroscopy showed that he had been infected with bacteria and had developed a stomach ulcer. It was immediately treated and cured on the advice of his wife. As a result of these so-called human clinical trials, he was eventually able to obtain funding from the National Medical Research Council to conduct a larger double-blind clinical trial to test the effectiveness of antibiotics in treating peptic ulcers. In 2005, he was awarded the Nobel Prize in Physiology or Medicine for his contributions to the field. In 2021, the US Department of Health and Human Services classified Helicobacter pylori as a carcinogen.

So why are tumor cells so vicious? It has to do with their temperament. Just as a person who is subjected to a series of unbearable mental blows will undergo a major change in temperament, so it is with tumor cells. Once the normal cells become tumor cells, they will "become monsters" and not go out of the ordinary. First, they will grow at an unprecedented rate and take over the place of normal cells, leading to the loss of normal cell function and affecting the function of tissues or organs. This is like a carpenter's nest being occupied by a dove. Second, due to their rapid growth, tumor cells require a large supply of nutrients and oxygen. However, with limited resources, they are forced to take away the supply from normal cells, leading to their death. Therefore, some people try to find ways to cut off the supply of resources to tumor cells, such as cutting off blood vessels that supply nutrients and lowering the level of oxygen. They do this in the hope of achieving the goal of starving or suffocating the tumor cells. However, the hope is always good, but the reality is cruel. Under conditions of starvation, tumor cells may stop growing indefinitely and go into a dormant state. However, this is just an appearance. Having closed their eyes and gone into a state of false sleep, the tumor cells quickly devised two counter-measures. The first is to break the arm to survive by letting most of the tumor cells die, reducing the consumption of resources, getting a small portion of the tumor cells to survive, and letting the living cells into a state of complete silence, almost invulnerable to any attack. Let the east and west, south and north winds blow; they are all indifferent. Once one day you are tired and let down your vig-ilance, they will come back. It can be said to be the weeds once burned by forest fire. No matter how dreadful the situation is, as long as the spring breeze blows, they will be reborn. Second, some tumor cells will wait for the opportunity to break through under strong killing pressure and move to other places to avoid the disaster of war. Once they have moved out, they will settle down and lurk in other strongholds to build up momentum for future counterattacks. Such counterattacks tend to be unpredictable and fatal. It can be seen that tumor cells not only have strong bodies but also extremely cunning minds. That is why, even after many centuries, tumor cells still threaten our lifeline.

In fact, tumor cells that are not killed acquire new properties that the pre-vious cells did not have, namely stem cell properties. We call these cells cancer stem cells. However, these stem cells are not the stem cells we mentioned earlier that promote tissue or organ regeneration; these stem cells are the devil incar-nate. Attributed to cancer stem cells, even if there is only one such cell left in the body, in the right environment it will start to replicate, expand and differen-tiate just like a normal stem cell, building up a huge army of tumor cells. Why do we often say that tumors are difficult to treat? One of the reasons is that there are too many cell types and there are so many different classes of cells that existing drugs can only kill one type of tumor cell and not others. Even if the children and grandchildren of tumor cells are killed, the ancestral generation of tumor cells immediately produces new progeny to replenish them, leading to treatment failure.

Cancer stem cells were discovered in leukemia as early as 1969, hence given the name "leukemia stem cells." However, due to technological limitations and the slow development of the stem cell field, it was not until 2003 that cancer stem cells were discovered in solid tumors other than blood cancers. John Dick in Canada was the first to identify such cells in breast cancer using flow cell sorting. In this technique, a group of cells is lined up and passed through a gate one by one. A laser is used to characterize each cell as it passes through. If different or odd-looking cells are detected, they are marked with a positive or negative charge. The aim is to isolate or enrich specific cells. Since then, researchers have discovered cancer stem cells in other solid tumor tissues, such as gliomas. Compared with other glioma cells, glioma stem cells have strong resistance to conventional chemotherapy and radiotherapy. It is difficult to kill them, and if a larger dose is used to achieve the purpose of killing them, it will often have a fatal effect on normal tissue. This results in the saying "killing most of one's enemies while losing oneself." Therefore, targeted therapy of tumor stem cells has become one of the important research goals of first-line oncologists. However, it is hard to say whether it can ultimately benefit patients. After all, the tumor stem cell theory is only one of many theories in the field of tumor research.

Are all tumor cells so stubborn that they are incurable? The answer is both no and yes. After in-depth studies, researchers have fully understood the nature of some tumors and can already achieve a cure, but for the vast majority of tumors, we still cannot do anything. So some doctors have proposed to live with the tumor to see the tumor as a kind of chronic disease. As long as it is controlled not to further develop and deteriorate and does not endanger life, then just let it go. Although it is a negative therapy, it can often give tumor patients a better quality of life.

As for cancer stem cells, one of the methods of treating them is differentiation therapy. The most successful case is the use of trans-retinoic acid (TRA) by Chinese scholar Academician Wang Zhenyi. He induced leukemia cells to turn back into normal cells, which he also called reformation therapy. Wang Zhenyi was born in Shanghai on November 30, 1924, and had seven siblings. His family was fortunate enough to provide their children good education even in those turbulent times, and he was no exception. At his father's suggestion, Wang Zhenyi entered Aurora University at the age of 18 to study medicine. Six years later, he received his doctorate in medicine directly and stayed on at Guangci Hospital, which is now Ruijin Hospital, Shanghai Jiaotong University School of Medicine, where he works until today. Afterward, under the guidance and leadership of renowned internal medicine specialist Kuang Ankun, he began to work with hematologic diseases, fully dedicating himself to the research and treatment of these conditions. In 1953, he joined the medical team of the Anti-US Aid for North Korea, and as a senior doctor of the Northeast Military Region Internal Medicine Circuit Medical Team, he treated the wounded volunteer army. He then focused on research into thrombosis and hemostasis and was the first to establish a diagnostic method for hemophilia in China.

Wang Zhenyi

At the same time, under the encouragement of the "Great Leap Forward" slogan in China at that time, Wang Zhenyi began to make contact with leukemia and put forward the slogan of conquering leukemia as soon as possible. However, slogans could not solve any problems, and he was saddened and blamed himself for the death of every leukemia patient before him. So long as there was progress in the international arena, he would pay attention and follow it up. In the 1970s, he learned that Israeli scientists had confirmed in animal experiments that leukemia cells could be differentiated into normal cells. In the early 1980s, he learned that American scientists had induced the differentiation of leukemia cells using a type of drug called cis-retinoic acid. In 1985, a five-year-old girl named Jingjing was admitted to Shanghai Children's Hospital where his wife Xie Jingxiong worked. Jingjing showed typical symptoms of leukemia, including high fever, nosebleeds, and perianal abscesses. She would have lost her life at any time if she hadn't received effective treatment. At that time, as the President of the Shanghai Second Medical College and already in his sixties, Wang Zhenyi suggested Xie Jingxiong to try treatment with TRA. However, TRA had not yet undergone any relevant clinical trials, and doctors would have to take huge risks if the treatment failed. With the patient's life on the line, they decided to have a try. Why TRA but not cis-retinoic acid as reported abroad? First, it was difficult to buy cis-retinoic acid in Shanghai at that time, and it would be extremely expensive to import it from abroad. Second, the Shanghai Sixth Pharmaceutical Factory happened to be producing TRA at that time, so he used it by a fluke and found that the latter had a better effect in inducing the differentiation of leukemia cells in the in vitro test. This time, the miracle not only favored Jingjing but also opened the door to a cure that Wang Zhenyi had always dreamed of. Soon, they were testing the drug on more leukemia patients, and almost 90% of them were cured. In 1988, they published a summary of their results in the leading academic journal in the field of hematology. The paper has been cited more than 2,000 times in the journal *Blood* and he has won numerous awards, including the National Top Science and Technology Prize.

It would be underestimating Master Wang if the story ended here. He has not only achieved success in his professional field but has also made remarkable achievements in the cultivation of talents. Three of his students, Chen Zhu, Chen Saijuan, and Chen Guoqiang, have been awarded the title of Academician. Each of them has made remarkable achievements in the field of leukemia, thus creating a great history of four Academicians from one mentorship lineage. On the basis of the positive therapeutic results achieved in the early stages, some of his best students, together with Zhang Tingdong of Harbin Medical University, further combined TRA and arsenic to raise the cure rate of acute promyelocytic leukemia to a new level. They turned a terminal disease into a curable one, and this new treatment protocol was internationally recognized by experts and became known as the Shanghai Protocol. Although it was not clear at the time what the specific reasons were for the role of these drugs, it was a great accomplishment of achieving the purpose of curing the sick and saving people. Through modern molecular biology technology, researchers have mastered the main causes of this curable leukemia and the mechanism of the combined effect of drugs. This has opened the curtain of precision treatment in the domestic medical field.

In fact, the first drug to achieve precision treatment in the international arena is imatinib. Its birth process can be said to be full of bumps and bruises. A few years ago, a film called *Dying to Survive* was released in China and it was very well received. It tells the story of patients seeking imatinib, a powerful anti-tumor drug. The story of the developer behind the drug deserves everyone's attention. If someone makes a film of this story, it will surely be a masterpiece spanning half a century and involving many unknown contributors.

Peter Nowell was born on February 8, 1928, in Philadelphia, the US. After graduating from the University of Pennsylvania with a degree in medicine, he served two years in the army before returning to his alma mater to pursue a career in pathology. At that time, the main research tool was the optical microscope, and it was used to observe whether the chromosome changes in different cells during cell division were abnormal. This improvement in methodology allowed him and his colleagues to observe the first chromosomal abnormalities in leukemia cells in 1960, particularly a shortening of chromosome 22, which was observed in almost all blood cells of these patients. This finding was never reported at the time, partly because the focus of research at the time was on tumors caused by viral infections, and partly because similar chromosomal variations were not found in almost any other tumor cells. Therefore, this chromosome was not given enough attention at the time and was simply called the Philadelphia chromosome.

Peter Nowell

Janet Rowley

Nicholas Lydon

It wasn't until 12 years later that Janet Rowley at the University of Chicago, who had further improved techniques for preparing and imaging chromosomes in cells, observed an exchange between this shortened chromosome and chromosome 9. She also found other mutated chromosomes in leukemia, establishing a link between chromosomal translocations and cancer. It took another ten years to understand the exact pathogenesis of the leukemia caused by the translocation between chromosomes 22 and 9. In this process, two different genes that did not go together fused, overactivating an enzyme that should not have been activated in the first place.

Now that a clear target is known, the next step is to inhibit the activity of the enzyme. For those in chemistry and pharmacology, the best way to do this is to use mass screening to find that specific inhibitor from thousands of compounds, then hit the bull's eye and get the job done. A few years later, Nicholas Lydon, a researcher at a small company, was lucky enough to find this inhibitor. He named it imatinib. However, the company's management at the time had no interest in taking the compound into clinical trials, so imatinib was left to languish for several years. It wasn't until the companies merged to form today's pharmaceutical giant Novartis that things took a turn. With continued lobbying by Lydon, the project continued and clinical trials were initiated in collaboration with oncologists Brian Druker and Charles Sawyers. Phase I trials showed excellent therapeutic efficacy and very few side effects. Imatinib was rapidly developed and brought to market at the beginning of the twenty-first century.

Since tumor cells are so repulsive, are they completely useless? From a dialectical point of view, we can make the best use of tumor cells by taking advantage of their unlimited growth. One of their contributions is the production of monoclonal antibodies. This is achieved by fusing tumor cells with antibody-producing lymphocytes to create a mono antibody.

César Milstein was born on October 8, 1927, in Puerto Blanca, Argentina, into a Jewish immigrant family that, like other Jewish families around the world, placed a high value on education. Even in those difficult times, all three of the youngest boys in the family graduated from college. Milstein was the second youngest in his family, and although he was an average student, he enjoyed participating in student government activities. The first thing he did after graduating was to marry his college sweetheart and then spend a year traveling around Europe. Upon his return to Argentina, he began studying for a PhD in medicine, focusing on enzyme kinetics. Due to a lack of stable financial support, he was only able to work and study at the same time. During this time, he traveled to the University of Cambridge in the UK on a scholarship to continue his research in enzymology, where he met the famous Frederick Sanger. After completing his PhD, Milstein returned to his home country for two years of independent research. However, due to the political climate of the time, he returned to England and joined Sanger's laboratory. There, he was advised to change his research direction from enzymology to immunology. It was at this time that it was discovered that not only blood plasma cells, which differentiate from B-cells in the fluid, could produce antibodies. Furthermore, the destruction of plasma cells led to multiple myeloma. Although antibodies had already been discovered at the end of the nineteenth century, these are substances that neutralize toxins when produced in the body. However, it took 30 years to realize that these substances are proteins in nature and that an antibody can only bind to one toxin, just as a key can only open paired locks. These toxins were later called antigens. Antibodies can be used not only to neutralize the toxins, but also to recognize the influenza virus and induce resistance. However, it was very difficult to obtain antibodies against just one virus because the antibodies secreted by normal plasma cells are a mixed group, like a well-made cocktail, and it was very difficult to isolate one of the components. The problem Milstein was trying to solve was how to obtain a large number of antibodies. The strategy he adopted was the then popular cell fusion technology, but unfortunately, it was not very effective.

The turning point came in 1974 when a recent PhD graduate, Georges Jean Franz Köhler, joined his laboratory for postdoctoral training. Born in Munich, Germany, on April 17, 1946, Köhler was an average student. He then went to a regular university to study biology, which he fell in love with. During his PhD, he worked in immunology at the newly established Basel Institute for Immunology in Switzerland. When Köhler took over, he introduced the technique of screening red blood cells for antibodies, invented by Niels Jerne, the director of the institute. He also fused plasma cells capable of secreting a single antibody with myeloma cells, which can proliferate indefinitely, to form a hybrid tumor cell. If this fusion cell was able to secrete a single antibody to kill the red blood cells, then the experiment was successful. So he was extremely nervous about the results and convinced his wife to observe them with him. When they saw hemolytic plaques in the red blood cells, they knew their experiment had succeeded and the first hybridoma cell that could secrete a single antibody and grow indefinitely was born. He was 29 at the time. He returned to his former institute and began to work independently. However, in a laboratory accident, he suffered from inhalation of smoke and died at the age of 49. The generation of hybridoma technology has greatly advanced the development of medicine and the life sciences, offering hope for the treatment of many clinical diseases. Köhler's life was short but brilliant. Jerne, Milstein, and Köhler were awarded the Nobel Prize in Physiology or Medicine in 1984.

Of course, hybridoma technology is only one of the contributions of tumor cells to humankind. In the next chapter, we will describe another great contribution of tumor cells.

Immortality of
Tumor Cells

$\mathcal{I}$f you ask what are the most famous events in the history of cells, there are many, but if you ask what is the most famous cell in the history of cells, it can only be the HeLa cells. HeLa cells not only appear frequently on the *New York Times* and other well-known media platforms but also have a number of books dedicated to them. The most representative of these books is *The Immortal Life of Henrietta Lacks*, which has also been made into a film. Of course, in addition to the value of the HeLa cells themselves, people are more interested in the story behind the HeLa cells and the medical-ethical debate that has arisen from them.

Before discussing the HeLa cells, there are two things that need to be mentioned. One is the Hayflick limit, and the other is Carrel's chicken cardiomyocytes. Before Hayflick, the consensus in the field was that cells isolated from the body could divide and proliferate indefinitely when cultured in vitro. However, Hayflick discovered that almost all cells can only proliferate for a certain number of generations, after which they are essentially driven to death—the famous Hayflick limit. The discovery of this cellular law made people realize that not only is the lifespan of an individual organism limited, but so are the cells it contains. However, there are always exceptions. Some cells that can be expanded in vitro to reach the Hayflick limit can actually survive and grow indefinitely, thus realizing eternal life. The French scientist Carrel found that chicken cardiomyocytes isolated from the embryonic stage of a chicken could grow indefinitely in vitro. This discovery opened the door for humankind to obtain immortal cells. At the time, the whole of Britain, and indeed the world, marveled at this discovery. But fame does not equal scientific truth. Later studies showed that the discovery could not be repeated. From this, we can see that not all Nobel Prize results can stand the test of history. Therefore, people should not be superstitious about the Nobel Prize, whether they are scientists themselves or part of the general public. Of course, after Carrel, scientists soon created cells that can live forever. However, these cells are basically obtained from animals such as mice, hamsters, or monkeys, and human cells have never been able to achieve immortality. Therefore, the immortality of human cells was a goal for scientists at that time.

Henrietta Lacks   George Gey

The HeLa cells are called HeLa because they come from a Black woman named Henrietta Lacks. Lacks was born on August 1, 1921, in Roanoke, Virginia, and lived in Maryland, the US. As a child, she lived with her grandfather on a tobacco farm. Due to the hardships of life, Lacks had to get up at four o'clock every day to milk the cows, feed the chickens and horses, and pick up the tobacco piles. The nicotine from the tobacco often made her hands burn. Luckily, life wasn't lonely. Her cousin, who is five years older than her, was always there for her, whether she was working on the farm or playing. It wasn't an easy life for them, but they were childhood sweethearts. When the time came to talk about marriage, it was natural for them to come together. The year they were married coincided with the Japanese bombing of Pearl Harbor in the US. The outbreak of the war brought prosperity to the tobacco industry, and their work became busier as a result. After her marriage, she became a kind and loving wife and mother of several lovely children. As she grew older, however, Lacks often felt pain in her body as she was with her own children and watching the sunset with her closest cousins. This pain continued to torment her and caused her to lose weight with each passing day. After repeated warnings from her family, she promised to go to the hospital to have her health checked. All mothers are the same, and so are all poor people. Even in the United States, the first thing that comes to the mind of the poor is to take care of their children with the little financial income they have. And even if they are sick, they try not to go to the hospital, which is very similar to many families and mothers in the rural areas of other countries.

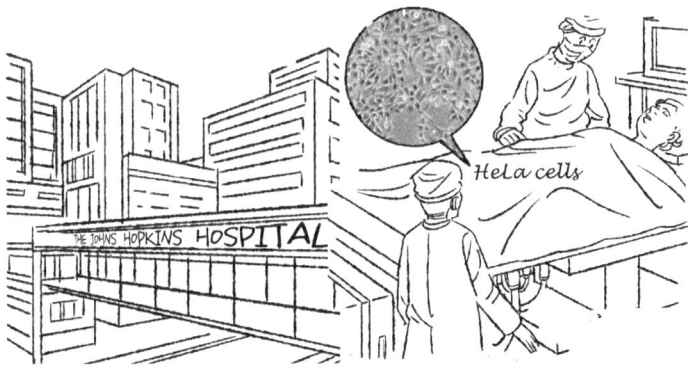

HeLa cells

THE JOHNS HOPKINS HOSPITAL

It wasn't until 1951 that Lacks, in unbearable pain, drove alone to the outpatient wing of the Johns Hopkins University Hospital after arranging for her family's living accommodations. The university is well-known in the United States and was founded with Hopkins's money, mainly to train dual-disciplinary doctors who are skilled in clinical medicine and basic medical research. Lacks was seen by a surgeon, and based on her test results, Lacks was definitively diagnosed with cervical cancer. She was soon scheduled for surgery, and samples of the surgically removed cervical cancer tissues were sent to Dr. George Gey's laboratory. Dr. Gey had been trying to isolate, expand, and share the cells from patients' tissue samples in the hope of obtaining infinitely expandable human cells. So whenever the hospital had a suitable clinical sample, he would bring it in to try. This time, he didn't expect too much from Lacks's cervical cancer sample and just followed the standard steps he'd taken before. He started digesting, isolating, culturing, passaging, and observing. Unfortunately, Lacks's condition did not go into remission as a result of the surgery, so the hospital continued to treat her with radiation. At the time, the technology was not as sophisticated as today's radiotherapy protocols, which use whole body radiotherapy or localized precision radiotherapy. Instead, a radioactive metal block was placed behind a special tube and inserted into the cervix for therapeutic purposes. Perhaps the treatment was not perfect, or perhaps the disease had worsened beyond repair, six months after her admission to the hospital, Lacks left her children and loved ones forever, away from the disease.

Although Lacks was gone, Dr. Gey's experiments did not stop. However, he observed a phenomenon that made him very happy. The cervical cancer cells isolated from Lacks' body were not only growing rapidly, but they were also able to reproduce continuously without the slightest desire to stop growing. Gey understood that he had finally cultivated a strain of human cells that could be immortalized, and that his experiment had been successful. In honor of the source of this cell strain, he took the first two letters of Lacks' surname and first name, respectively, naming it HeLa. The acquisition of the immortalized HeLa cells was not only a success for Gey and Johns Hopkins University, but also a glory for science. It soon went to industry, providing the world with a steady supply of human source cells for countless experiments.

The savvy pharmaceutical company quickly acquired the patent and rights to use the HeLa cells from Johns Hopkins University and quickly set up the first cell factory in human history. In the cell factories, huge agitated culture systems replaced lab flasks or Petri dishes. Thousands of liters of culture fluid were consumed, and thousands of tons of HeLa cells were produced and packaged into freezing tubes or culture flasks that could withstand long-distance transport. They are either given away to research institutions or sold to pharmaceutical companies for tens or hundreds of dollars. From hospitals to factories, from a small town to the entire United States, and to organizations large and small around the world, HeLa cells have made a magnificent transition.

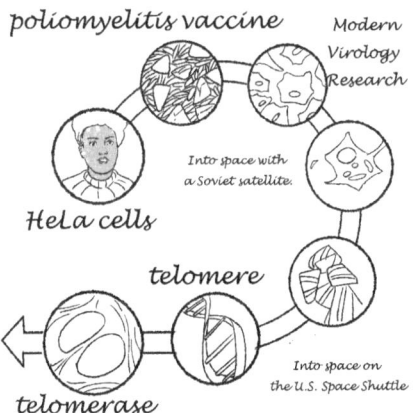

Using these immortalized HeLa cells, researchers immediately began a study of polio, for which there was no cure at the time, and achieved positive results. Poliomyelitis is an infectious disease caused by a viral infection that primarily affects the central nervous system. In severe cases, it can lead to paralysis. Most patients are children, which is why poliomyelitis is also known as the paralysis syndrome of children in China. As there is no effective treatment for polio, prevention is the only hope. But at the time, there was a lack of effective prevention, and children in the United States were suffering badly from polio. Although scientists had developed a vaccine against the virus, the lack of an effective means of validation at the time prevented its widespread introduction and use. Using these infected cells, it was confirmed that the vaccine that had already been produced not only had a better safety profile but also effectively inhibited the virus in the cells, which strengthened the resolve to promote the vaccine in the population. It is on the basis of mass vaccination, not only in the United States but around the world, that the threat of polio to humans has now been virtually eliminated. The HeLa cells are absolutely credible in this process. Immediately afterward, scientists tested the ability of other viruses to infect HeLa cells, such as the measles virus, the herpes virus, and the human papillomavirus. This marked the beginning of modern virology. In addition to viral research, HeLa cells also traveled into space on Soviet satellites and NASA space shuttles, becoming the first human cells to be flown into space to test the effects of the space environment on human cell growth. This laid the groundwork for the future use of the space environment to treat disease and for human migration into space.

In addition to this, another important property of the HeLa cells is that they provide important research material for humankind to explore the secret of immortality. Through the study, it was found that the important reason why such cells can continue to grow and cross the Hayflick limit lies in the activation of a special enzyme called telomerase in their bodies. Telomeres are important structures that protect the stability of chromosomes in the nucleus. They are like hats for the chromosomes, without which they would become shorter and shorter with each cell division. As a result, the life of the cell gradually dies as the chromosomes become shorter and shorter. The presence of telomerase prevents the telomeres from shortening and keeps the cell alive. Of course, with regard to the cellular turnover of old and new cells, as we discussed earlier, cell renewal according to physiological and developmental needs is necessary and beneficial to maintain the function of tissues or organs. If abnormal, it can lead to the development of tumors. Therefore, how we can balance cell death and immortality to better serve human regeneration or life extension needs to be further explored. There is still a long way to go, but the discovery of telomerase in HeLa cells provides a bright light to move forward.

Half a century later, HeLa cells are still contributing to modern scientific research. Thousands of papers are written about them every year, and the number is still growing. With the help of modern science and technology, we have finally found out why HeLa cells can be immortalized. This is mainly because Lacks herself was infected with the human papillomavirus, which causes cervical cancer. The virus inserted its oncogenes into the genes of normal cervical cells, leading to cancer transformation and immortalization of the cells.

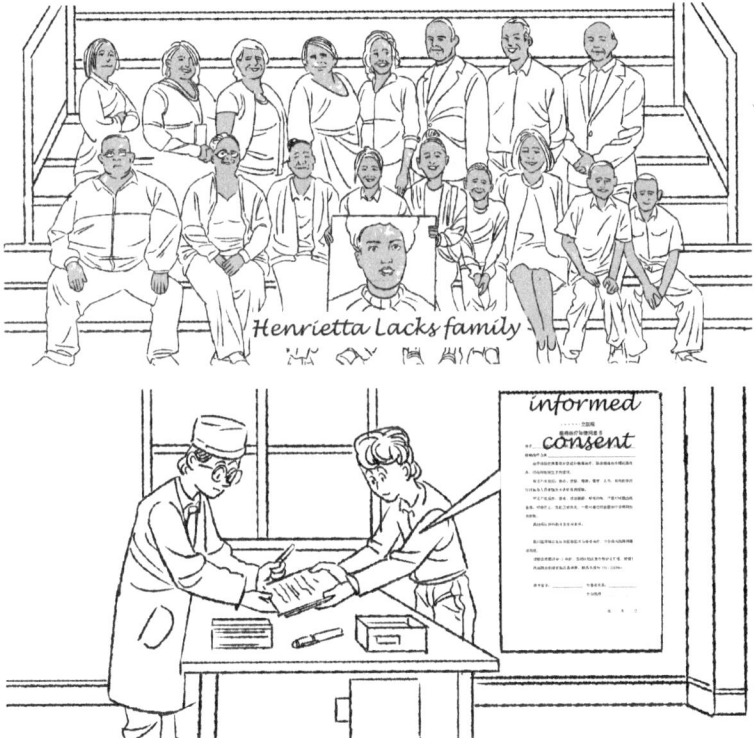

Immortalized HeLa cells changed the world of medicine, scientific research, and the fate of many people. Huge amounts of wealth flowed into the pockets of corporations and capitalists as HeLa cells expanded from generation to generation. But not a penny went to Lacks herself or her family, who didn't even know that their mother's or wife's cells were still alive in the world. It was only when researchers began to study the genetic properties of HeLa cells and asked Lacks's children to donate blood so that their family's genetic history could be analyzed that this thin layer of window paper was pierced, and Lacks's relatives began the arduous journey of defending their rights.

There are many kinds of interest in defending rights, perhaps money, perhaps honor, perhaps something else. For Lacks's children, perhaps more than anything else, they wanted to know if their mother is still alive. For an ordinary person, when a loved one dies and they are told that he or she is still alive, they must be filled with confusion and surprise. They were not even sure what a cell is. When they were told that only cells isolated from a tumor in their mother's body had survived, they took it as that their mother was still alive. Wearing thick gloves and holding the cryopreservation tubes of HeLa cells that had just been removed from liquid nitrogen, they watched the rising mist. They felt so warm, as if their sleeping mother was lying in the palm of their hands, as if she was a baby. The researchers developed fluorescent photographs of the HeLa cells, framed them and gave them to the family. Lacks's family hung the photographs on the walls of their homes, where their mother looked like shining stars in the sky, watching and accompanying them all the time. Gey then set up the HeLa Foundation, and at its first annual meeting, Lacks' descendants were invited to present her life. It was also a relief for them to know that their mother was still being honored and cared for by so many. In addition, the foundation built a memorial in Lacks's home town, displaying Lacks's memorabilia and photographs. Of course, all of the above happened after the legal defense, which is seen as a more positive side.

The other side of advocacy is often painful, and Lacks's loved ones will not get the answers they deserve. There was financial compensation, but the company had already made a lot of money and Johns Hopkins University had benefited from the patent transfer. The main reason Lacks' family did not receive financial compensation was that research ethics were not well developed at the time, and there was no signed informed consent for the use of clinical samples. The doctors tacitly approve that Lacks had consented to tissue donation for scientific research and possible commercial applications at a later date, not to mention it was a meaningless, discarded, disgusting piece of meat to herself and her family. Until 2020, it took almost 70 years for Lacks's family to receive the compensation they deserved. In fact, the biggest research ethics scandal in American health care system at the time was the "Tuskegee Syphilis Study." The patients, most of whom were African Americans, were not informed of their condition, nor were they provided effective treatment, even after penicillin became the standard cure for syphilis in the 1940s. Instead, researchers aimed to observe the natural progression of untreated syphilis, which led to many participants suffering severe health consequences, including death. To some extent, this scandal led to public distrust of the hospital and escalated tensions in the Black movement.

Even today, despite informed consent in hospitals, tissue samples from patients are still abused or used commercially. How to prevent similar tragedies from happening again may be a question that both clinicians and scientists ask from time to time. This applies to tissue samples, cells, biologically active substances, and, in particular, genetic information, which is not yet in the public domain but is already a matter of personal privacy. Today, countries have recognized the importance of protecting human genetic resources. International transfer seems to have been regulated, but the commercial exploitation of one's own biological resources and information still falls into a gray area that lacks regulation.

For thousands of years, tumors have been considered the enemy of humankind. Only a few types of tumor cells have been domesticated by us to serve the modern bio-industry. However, we do not yet know if tumor cells have other uses. As with the preservation of Earth's biodiversity, many organisms, both animal and plant, have quietly disappeared from the face of the Earth before we humans recognized their existence and the value of their importance. It is important to realize that since the beginning of life on Earth, all living things have interacted with each other and there is a certain balance. Although the disappearance of a species does not necessarily create a butterfly effect, it will be too late to regret if we want to recognize or take advantage of it one day. In the case of tumors arising from natural evolution, each tumor cell contains information about the spontaneous mutation of the cell in response to external stimuli or pressures, just as humans use artificial radiation or space radiation to mutate plant cells or seeds for the purpose of breeding. Therefore, the ability to mimic artificial breeding and extract useful information from billions of spontaneous mutations in tumor cells may be an important resource in the future. However, with our current level of understanding, we are not yet able to recognize its importance. Establishing a genetic resource bank of human tumor cells to create a permanent repository of these hidden secrets may be another way to achieve tumor cell immortality.

Because the immortalization of tumor cells plays such an important role, and because cells obtained from tissue isolation are more likely to acquire the ability of immortality compared to the normal group, people have established corresponding tumor cell lines for almost all types of tumors after the success of the establishment of HeLa cells. Of course, some of these immortalized tumor cells are directly isolated and screened from different tumor tissues, while others are genetically manipulated by using viruses to directly infect tumor cells in vitro, and then forcibly achieve the purpose of immortality. Among these viruses, the most commonly used is monkey vacuolar virus 40, which is an extremely powerful virus that can not only prompt the immortalization of tumor cells but also have the same strong effect on normal cells. These immortalized tumor cells not only provide viable raw material for extensive research into the biological properties of different tumor cells and the development of drugs and methods to inhibit tumor cell growth and induce tumor cell death but also provide new ideas for the development of cell therapies for tissue regeneration.

When we introduced the members of the blood cell family, we mentioned that erythrocytes in the blood have an important function in transporting oxygen. However, erythrocytes do not have a nucleus, so there is no way for humans to obtain more erythrocytes through erythrocyte expansion. At present, clinical practice mainly relies on volunteers to donate blood. The red cells are then separated and used for later component transfusions. Hemorrhage due to trauma often requires the transfusion of large quantities of red blood cells. Given the huge clinical demand for red blood cells, obtaining sufficiently large quantities has become a strategic need for a country. Inspired by the immortalization of tumor cells, scientists are trying to immortalize small precursor cells that can produce red blood cells. These cells are called erythroid progenitor cells and are also descendants of hematopoietic stem cells. However, they can only develop into red blood cells and not into other blood cells. Although erythroid progenitors have a nucleus and can divide and expand, they cannot escape the Hayflick limit, just as the Earth cannot escape the solar system. Therefore, the ability of these progenitors to produce erythrocytes is limited. If we immortalize these progenitor cells, such as the monkey vacuolar virus 40, we can establish immortal erythrocyte progenitor cell lines and further adopt an industrialized cell culture system. This will allow us to shift the procurement of erythrocytes from the limited method of blood donation to the unlimited production mode of a factory. This solution will solve the problem of erythrocyte sources for clinical treatment once and for all. At present, this method has made promising progress. If the safety of these manufactured red blood cells can be guaranteed, patients will surely benefit from these new technologies within the next few years.

Strange Creatures

*H*aving said so much about cells and the attractive therapeutic prospect of stem cells, but for ordinary readers cells are still something untouchable and invisible. So, is there a good way to make people really feel the existence of cells and the charm of stem cells? Here are some animals in nature that we can easily come into contact with and the amazing role of cells in their bodies.

gecko

In the summer heat, the wings of robins and flies sound endlessly during the day, as if everything is drowned in the squeak and buzz. In the evening, swarms of swallows and bats circle back and forth at low altitudes. At night, the most annoying sound is the call of blood-sucking mosquitoes by the ear. In addition to these animals that can be observed by the discerning eye, there is a group of small animals that are often active in the summer and are very close to us. They can even be seen on the walls of the house. This is the gecko. Though these four-legged snakes are small, they are capable of a lot. As well as eating mosquitoes and destroying pests in the house, they have a unique ability to break their tails in order to survive. Although they are very fast, they are too small to be easily hunted by other animals. To avoid being hunted, they will cut off their tails at critical moments. The disembodied tails will continue to bounce around, attracting the attention of natural enemies. They will use the hard-won brief moment to make a quick escape. Don't worry for them. It doesn't take long for a new one to grow back. Why can geckos grow new tails? And how do they do it? This is mainly due to the clear division of labor and mutual cooperation of different types of cells in their bodies, including immune cells, muscle cells, and nerve cells. In the stimulation of losing a tail, these cells will quickly join the group, rapidly proliferate and grow. They will perform their respective tasks in accordance with the negotiated strategy and a very tacit understanding. The ones can repair the muscle repair the muscle, the ones can repair the skin repair the skin, and the ones can repair the blood vessels repair the blood vessels. Although it could not be completed in one day, as time passed, the cells increased day by day, the tissues took shape day by day, and eventually grew into a complete, brand-new tail.

Except for geckos, there is another group of small animals that have the ability to regenerate their tails—the zebrafish. This kind of tropical fish have a body length of three centimeters. Their body, from the tail to the head, has black stripes in a pattern that is extremely similar to that of a zebra. Hence, they are called zebrafish. Unlike geckos, zebrafish's tails won't break off on their own. They only detach if the fish are attacked by other animals or suffer an injury. As a common pet fish, zebrafish are widely available, and you can keep them in a small aquarium at home. If you ever see a zebrafish with a damaged tail in your tank, don't worry—its tail can often grow back in time. Their regeneration depends on cells in the tail that multiply and grow after injury. However, if these important cells are too damaged or missing, in other words, the injury on the fish's tail is too close to its root, the tail may not fully regenerate. In addition to being able to regenerate its tail, the zebrafish also has another ability that the gecko does not have: the ability to repair and regenerate heart tissues. In both geckos and humans, once the heart is damaged, even tiny, localized cells are dead or missing, leading to serious symptoms. And once these cells are lost, there is no way to regenerate new ones. But zebrafish are different: if their hearts are damaged in a way that is not immediately fatal, as in the case of their magic tails, after a while the heart cells around the damaged area start to grow back and fill in the missing heart area, thus escaping a fatal death.

Animals with regenerative capacity in aquatic organisms are far more numerous than terrestrial organisms, both in terms of their capacity and in terms of their numbers. Compared to the zebrafish mentioned above, starfish have a better regenerative ability. In the spring of 2021, a strange incident in Qingdao, Shandong Province, China, was widely discussed. Clams and starfish were the main characters. Originally, clams were one of the most important seafood species cultivated by local fishermen. However, when it came time to harvest, fishermen found that when they cast their nets, they only saw starfish, not clams.

Of course, whether this is a blessing in disguise is a matter of opinion. Clams are considered delicious, whereas starfish are also an unusual delicacy. Why do starfish blooms occur? There are a number of possible reasons, including the fact that adult starfish are highly prolific and have few natural predators. Another important reason is that, as a sea creature with five tentacles shaped like a pentagram, their hard, mottled shell is dominated by reproductive glands. If a tentacle is lost, a whole new one will soon grow back. All five have the ability to regenerate. It is clear that the seemingly fragile little starfish have a hidden universe and are born with a strong ability to survive.

salamander

Salamanders are amphibians that live in mountain forests and streams. They are also simple and somewhat tissue-regenerative animals. Their ability to regenerate tissue is most evident in their limbs. When the front or back limbs are amputated, the remaining parts form a mass of fleshy tissue that soon turns into a small bud. Don't look down on this bud, which is undergoing more complex cellular activity than the regeneration of a gecko's tail. The bones in the limbs are essential for supporting the whole body. The feet at the ends of the limbs have a more specialized branching pattern, making regeneration a more demanding challenge. Inside the buds at the site of injury, all types of cells, including skeletal cells, grow and work together to form tissues of different lengths and precisely according to the original structure of the limb until the end of the toe is complete. Living in a city of steel forest, if one day look at how all those skyscrapers are going up, you will find that in the process of growth of tall buildings, the top will be wrapped in a big machine. With the rise of the building, this machine will also be lifted like a hat, until the final capping of the completion. This process is like the regeneration of limbs in a salamander.

If the salamanders have dazzled you with their skills, the next little guy to be introduced will definitely make your jaw drop. It's said that after their bodies are chopped up into mincemeat, each piece of mincemeat can turn into a complete body. Isn't that incredible? A bit like the Terminator? And these little guys are definitely not creatures from the future, but planarians that have been living on Earth for tens of thousands of years. Planarians look small, with a triangular head that has two eyes the size of a rapeseed, a flat white body that is just two or three centimeters long, and a small tail that is not too long and not too short. How did people discover the incredible regenerative abilities of planarians? It all started with some fascinating experiments. For example, when a planarian is cut in half, instead of dying, something amazing happens. Within a few days, each half can grow into a complete planarian, as if they now have two heads! But that's just the beginning. If a planarian is cut into many pieces, you might think this would be the end for it, right? Surprisingly, each part—whether it's from the head, body, or tail—will regenerate and become a whole, living planarian. Of course, planarians' super-regenerative ability has yet to be thoroughly researched. It is known that each cell of planarians has greater plasticity than human cells. Once encounter a damaged stimulus, these cells will quickly enter a new state of formation, no longer just play before the local organization of cellular function, but follow the command where there is a need to where, as special forces in a variety of skills, can be engineers, can be artillery, can dive, can fly a plane, but also can carry out network security, as long as there is a need to play a role in where. It is this special nature of the planarian cells that determines the worm has a strong regenerative capacity.

The planarians have already shown amazing regenerative powers, but it is still very deadly when it sheds. However, if you doubt that animals in nature are capable of such things, you are mistaken. Nature's ability to create is far beyond our imagination. We always say that no matter how strong you are, there's always someone stronger. And when you think something is incredible, there is something even more incredible waiting to be discovered. So, after talking about planarians, we have to talk about sponges, a kind of creature that can actually be resurrected after being broken into pieces. Friends who like to watch the cartoon *SpongeBob SquarePants* will be familiar with sponges, and we are talking about sponges in the ocean. Sponges are a type of creature that live in the sea. When they were young, they were like a willow fluttering in the wind, and when they grow up, they are really like the sponges at home that wash the dishes. They normally feed on microorganisms in the water by constantly inhaling and filtering seawater. Even if a pile of sponges is broken into individual cells by an external force that is even more crushing than breaking them into pieces, each cell will voluntarily come back together to form a new baby sponge. No matter how many man-made mechanical methods are used to ravage it, even with a homogenizer to a paste, at the end there will still be a living and beautiful baby sponge. This is the sponge that is said to have the strongest regenerative capacity known on earth so far. Its secret lies in the special sponge cells. Although the sponge itself is a multicellular organism, its cells seem to have the characteristics of single-cell organisms. That is, each cell is like an independent individual life.

What is a single-cell organism? The species of animals mentioned above are basically multicellular organisms, which means that these animals are made up of several cells. Different cells perform their own unique functions, and the different functions of the cells work together and complement each other to support an independent living organism. Single-cell animals, as the name suggests, have only one cell. One cell represents one individual. The most representative of these animals is the paramecium. It is a small animal, only tens to hundreds of micrometers long from head to tail. The width is equivalent to two to ten strands of hair put together. Because its body shape resembles the sole of a shoe, with many cilia protruding from the edges, just like the straw shoes people used to wear, it is called the slipper-shaped bug in China. Because it consists of only one cell, the paramecium does not distinguish between male and female, but a hermaphrodite. Whereas we used to compare the cytoplasm to the belly of a cell, in the case of the paramecium, we can say that the cytoplasm is the belly. Unlike the belly of an animal cell, the paramecium's belly has two nuclei: one large and one small. On one side of the body, there is a small indented opening, which is the mouth. It feeds mainly on bacteria in the water. A paramecium can probably eat almost 2,000 bacteria an hour and can destroy more than 40,000 bacteria in a day. Into the stomach, the bacteria form a food bubble of 30 bacteria, which is slowly digested in the body. When this is done, it is expelled through a small hole at one end of the body. As they have no limbs, paramecium rely mainly on the wiggling of cilia around their bodies to move, allowing them to swim freely in the water. These little guys usually live in rice fields and small ditches, so they are also considered to be little environmental protectors in the water.

Single-cell organisms in nature are not only known as paramecium, but also as the infamous brain-eating bugs. The bacteria that paramecium love to eat are actually unicellular organisms. Although the latter are often classified as microorganisms, they are in the same class as bacteria, mycoplasmas, and fungi, among others. Let's take a look at each of these heavyweights.

Brain-eating bugs are just what they sound like. They like to eat the brain tissue of animals, which leads to meningitis and necrosis of the brain tissue and even death of the infected organism. The first human case was reported in Australia in the 1960s, and since then, one or two cases have been found every few years. The United States has reported the most cases to date. Although it sounds very scary, the incidence rate is extremely low and it's a rare disease, so we don't need to worry too much. So, what exactly is a brain-eating bug and why is it so dangerous? A brain-eating bug is an amoeba protozoan that has only one cell in its entire body and is so varied in shape that it is also known as an amoeba. It is about the size of a paramecium, but has a transparent body with a nucleus and cytoplasm in its belly. They usually live in damp soil, rivers, lakes, and stagnant water, which includes swimming pools that have not been sterilized for a long time and water from water pipes. For those who like to swim in the wild, the amoeba will enter the body through the nostrils because it has neurophilic characteristics. Once inside the body, it will crawl along the nerves to the brain tissue, causing serious nerve infections. If not treated in time, it will result in death.

When it comes to bacteria, we must not be unfamiliar with them, especially in the summer. A little carelessness can lead to diarrhea, which is caused by bacteria. The most famous type of bacteria is Escherichia coli (commonly known as "E. coli"). Bacteria are everywhere, not only in our neighborhood but also on and in our bodies. Most of the cells Leeuwenhoek observed with his homemade microscope were bacteria in water. The word "bacteria," however, was coined by the German Ehrenberg in 1828 and comes from the Greek word that originally meant a small rod. As well as rods, bacteria come in a variety of shapes, including spheres and spirals, and each type is very different from the other. Although the shapes vary greatly, the overall body structure of bacteria is quite similar. Unlike the cells described above, bacteria do not have a nucleus in their belly and the genetic material is exposed in the cytoplasm. However, there is a complexly structured cell wall on the outside of the cell membrane. Don't underestimate this cell wall, which is different from that of previous cells. Whether the bacteria want to capture other cells or survive in harsh environments, they have to rely on it. It is because of its powerful ability, to some extent, led to the number of bacteria being uncountable. They live in places you might never expect—they seem to be able to thrive almost anywhere! Whether it is at room temperature or freezing cold North and South polar ice and hot as fire in the magma, there are bacteria present. Generally speaking, everyone seems to be reluctant to talk about bacteria. That's because bacteria are often seen as bad guys that can lead to a variety of sensory infections and even be fatal. For a long time, the discovery of penicillin was considered one of the greatest discoveries in the history of human medicine. It has cured a wide range of diseases caused by bacteria, and its main mechanism is by destroying the cell walls of bacteria. But the truth is that just as there are good and bad people, normal and malignant cells, so do bacteria. For our bodies, whether it is the flora on the surface of the body or the flora in the gastrointestinal tract, the vast majority of them protect or support our health and are in a symbiotic state with us. Not only that, but the diseases caused by dysbiosis can be treated by eating the missing bacteria.

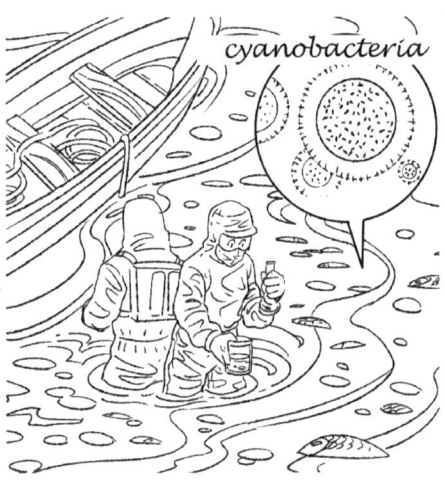

A type of bacteria called cyanobacteria is also known as blue-green algae because it is classified as a bacterium, but it is more similar to an alga. Cyanobacteria are a large family, often up to 10 microns in size, and up to 70 microns in the case of trematodes. Although they are unicellular individuals, some like to live in groups, forming clusters or filamentous structures, such as Candida and collar algae. In summer, many ponds, ditches, and even polluted rivers produce large amounts of algae. Examples include Taihu Lake in Jiangsu Province, China, and Dianchi Lake in Kunming, Yunnan Province, China, where algae blooms, including cyanobacteria, have occurred. Slightly different from bacteria, the outside of the cyanobacterial cell wall has an additional layer of structures called sheaths. These are made up of mainly acidic substances such as sugars and pectin and are essential for resisting the harsher external environment. Cyanobacteria have chlorophyll and cyanophyllin in their bodies, which gives them their blue-green color. Of course, there are also some cyanobacteria that contain other pigments in their bodies, such as red and yellow pigments, which give them their corresponding colors. When it comes to chlorophyll, many people know that it is the site of photosynthesis in plants and works in a similar way for cyanobacteria. Although some bacteria can also use sunlight for photosynthesis, they can only produce organic matter and cannot give off oxygen. In contrast, photosynthesis in cyanobacteria is completely modeled on that of plant cells, synthesizing nutrients as well as releasing oxygen. In this respect, cyanobacteria are closer to plant cells. However, the former are unicellular organisms that can move, while the latter are multicellular organisms that cannot.

Like bacteria, mycoplasma is a single-celled organism without a nucleus. It can be a nightmare when it appears in animal cell culture in vitro. Mycoplasma is also similar in size to bacteria, and although it is mostly round in shape, it has no cell wall and only a cell membrane. As a result, its body shape is usually very deformed, and it is far less infectious than bacteria. It usually prefers to infect the cells of the human urinary and reproductive systems, leading to diseases such as urethritis and cervicitis. Mycoplasma was discovered in 1989. Antibiotics that destroy cell walls, such as penicillin, are not effective against mycoplasmas because they do not have cell walls. But it's not like there's no way to deal with them at all. Other antibiotics, such as erythromycin, streptomycin, and tetracycline, which do not destroy the cell wall but rather the cell membrane and the proteins on the membrane, are still highly lethal. We may not necessarily understand these antibiotics, but for the last one, you must have heard of, especially with the infamous tetracycline teeth, which is caused by long-term use of this antibiotic, leading to pigment deposition on the teeth and resulting in yellowing.

Next, let's talk about yeast. It is inextricably linked to our lives, whether it is the steamed bread we all love to eat or the beer we love to drink. Yeast is indispensable. Yeast belongs to the fungi family, so what are fungi? To put it simply, it is the mushroom family, and its early name comes from the Latin word for mushroom. Of course, there are many members of this family, and yeast is just one of the oval-looking branches that also includes molds. In this case, the filamentous molds found in spoiled food are included. Fungi have a cell wall outside the cell membrane, but its composition is different from that of bacteria and more similar to that of plant cells. We will introduce plant cells in the next chapter, which are mainly composed of chitin and cellulose. Fungi not only have a nucleus in their belly, but also other common organelles such as mitochondria, endoplasmic reticulum, and lysosomes. Although fungi have a cell wall, their singular composition makes them rarely pathogenic. Instead, they play an important role in our daily lives. In addition to the previously mentioned bread and beer, yeasts are also involved in a variety of food production. From Pixian Broad Bean Paste of Sichuan to Jinhua Ham of Zhejiang in China, they all have a role to play. It is these mushrooms that finally allow for the creation of a variety of human delicacies, catering to the taste buds of gourmets.

The organisms mentioned above, whether multicellular or unicellular, are all living individuals with cellular forms. There is another kind of life form on Earth, which was born almost at the beginning of the Earth and has accompanied humankind for hundreds of millions of years. However, it does not have the typical cellular form. Strictly speaking, it can only be regarded as a simple combination of some organic substances, which is the virus. Viruses are very small, so small that they cannot be observed with our conventional optical microscope. We have to resort to the more powerful electron microscope to see their blurred image and existence. But when they are enlarged to a size that can be seen with the naked eye, we find that they have a variety of strange shapes. Some viruses look like machines, while others resemble alien spaceships in science fiction movies. They have extremely symmetrical geometric structures that even the best mathematicians would have a hard time imagining and constructing out of thin air. These viruses are made up of proteins and nucleic acids. Some viruses are made up of a simple piece of nucleic acid, which we call a virus-like substance, and others are made up of a small, insignificant piece of protein, which we call a prion. Viruses are small and depend on cells for their survival and reproduction, but overall they are like demons unleashed from Pandora's box. If bacteria can be said to have both good and bad qualities, viruses can only do more harm than good to humans. They cause various infectious diseases, overwhelming plagues, localized infections, and even create deadly tumors. Especially since we have just experienced the global COVID-19 pandemic, we have nothing but hate for viruses. However, the new coronavirus is only a minor player in causing human disease. More deadly viruses, such as Ebola, hide in the shadows, peering at us and leaving us defenseless.

To study these deadly little guys, scientists need to further increase the safety level of cell culture rooms. Currently, the safety level of cell culture rooms is internationally categorized into four levels, from one to four, with the higher the number, the higher the safety requirements. Laboratories that routinely perform animal cell culture have a level two, and if it involves the culture of viruses with a certain intensity of infectivity, such as COVID-19, it has a level of at least three, with the highest level for laboratories that culture the Ebola virus. At the highest level, all objects and gases entering and leaving the room must be completely filtered and sterilized. Otherwise, a silent catastrophe is inevitable.

Plant Cells Go Crazy

*H*aving introduced the history of animal cells, it is now time to focus on plant cells, which make up almost half of the planet. As described in the first chapter of this book, the discovery of cells began with the observation of plants. Overall, there is not much difference between plant and animal cells. Whether it is the cell membrane, cytoplasm, nucleus, or organelles, they are basically similar. The biggest difference is that plant cells have an extra layer of cell wall outside the cell membrane. In fact, the cell structure observed by Hooke is this cell wall structure. The cell wall is like a brick wall that can hold the cell firmly in place and limit its movement, which is considered to be the main characteristic difference between animal and plant cells. It also explains why animals can move on their own, whereas plants are immobile or at least not actively moving.

So, what is a cell wall? Like the extracellular matrix of animal cells, the cell wall is the extracellular matrix of plant cells. However, it is made up mainly of cellulose, which forms the protective shell of the plant cell. The most direct way to feel cellulose is to eat old celery or lettuce. The material that often gets stuck in your teeth is tiny fibers of cellulose, aggregated and linked together to form

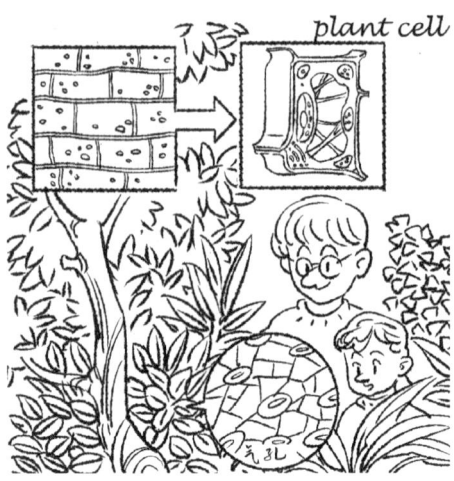

visible fibers. It is these fibers that give the plant its upright, weather-resistant stature and also determine its inactive fate. While it is only a support for the plant itself and is disliked by humans because it gets stuck in their teeth, it is an excellent food for herbivores. So, what is cellulose? It is essentially a long chain of tiny sugar molecules linked together, mainly glucose, with small amounts of galactose and xylose. So, for herbivores, once ingested, the cellulose is digested into individual sugar molecules. These are the substances that can provide energy. As for whether these animals can taste the sweetness, it is thought that they cannot because digestion takes place in the stomach, not in the mouth. Therefore, the tongue, which senses the deliciousness, is not able to enjoy it in time. However, ruminants such as cows and camels may be able to experience a hint of sweetness.

So, what role does the cell wall play in shaping the shape of plant cells, and does it limit cellular changes so that plant cells lack the variety of animal cells? The cell wall is indeed a master craftsman with the power of the devil. As a line shaper, it gives the plant cell an angular and rigid character. When the cell is cut from it, most of the cut surfaces show polygons, squares, pentagons, hexagons, and more. The three predominant structures of the plant, from bottom to top, are the root, the stem, and the leaves. There are both similarities and great differences between the cellular forms of each part. Root cells have thick and thin cell walls, while the cells in the stem are more elongated and interconnected. They often form extremely thin conduits that carry the water absorbed by the roots upward in a continuous stream. Another specialized organelle in the belly of the plant cell, the vesicle, plays a role not only in storing water but also in giving the cells a variety of colors. Because of the colorful pigments stored in the vesicles, the cells in the leaves take on a fascinating hue in autumn. Countless artists have left us treasures of paintings related to this, such as Vincent van Gogh's *Autumn Landscape*, Nikolai Pozdneev's *Autumn Day*, and Thomas Cole's *Distant View of Niagara Falls*, to name but a few. At this point, you may ask, "Is the most common green color in leaves also due to pigmentation in the vesicles?" The answer to this question will be revealed in later paragraphs. Another thing that is rather peculiar about the cells in a leaf is their arrangement. If you pick up a leaf and do not look at it carefully, you will probably think that it is a structure without any empty space. In fact, if you stare at it with your eyes wide open, get closer to it, or look at it with a magnifying glass, you will see that it has many small holes, just like the pores on our skin. These holes are the channels through which leaves breathe. Each one is made up of several cells in a circle, wrapped around two cells that are arched over each other and face each other, forming a gap. If you put your palms together and then press them against each other, you get an enlarged version of the leaf's pores.

There is one question that almost everyone likes to ask about plant cells, and that is about pollen. As with animal cells, everyone likes to ask: is an egg a cell? Pollen is similar to male reproductive cells in animals, and in different plants, pollen has different structures which can be exquisite when viewed under a microscope. The inside of each pollen grain is made up of two main cells: a reproductive cell that produces offspring and a nutritional cell that provides nutrients. The latter is often wrapped around the former.

Another important feature that distinguishes plant cells from animal cells is that they contain chloroplasts, which give leaves their green color. As one of the most important organelles in the cytoplasm of the cell, the main function of chloroplasts is photosynthesis. This process converts light into energy that the cell can use, while at the same time converting water and CO_2 in the air into nutrients needed for plant cell growth and releasing oxygen. Don't underestimate what is contained in the simple sentence above! It took more than two full centuries of human observation and effort to reach this conclusion.

In 1771, Joseph Priestley of the United Kingdom discovered that lighting a candle in a closed glass jar would consume the air in the jar, causing the candle to go out and the mice inside to die. However, if a plant was placed in the jar, it would not die and would flourish under the right conditions. In later studies, he discovered that the life-sustaining gas was oxygen released by the plant. In 1779, Jan Ingenhousz from the Netherlands further discovered that the plant in Priestley's experiment depended on two conditions: light and its green leaves. Since then, discoveries have continued to show that plants can use light to convert substances such as water and CO_2 into organic matter. However, the mechanism by which this happens was still not understood. It was not until the early nineteenth century that German scientist Richard Willstätter discovered the important role of chlorophyll. Willstätter was born in Baden, Germany, on August 13, 1872. At the age of 18, he entered the University of Munich, where he studied chemistry under Nobel Prize-winner Bayer and stayed for 15 years, progressing from student to lecturer. During this time, his research focused on alkaloids in plants, including their structure and synthesis processes. The infamous cocaine is a member of the alkaloid family. After more than ten years of experience, he felt that these studies were too superficial and wanted to take on some difficult research projects, such as pigmentation in plants. At the age of 33, he jumped ship and moved to the Swiss Federal Institute of Technology in Zurich to study chemical methods. Seven years passed in a blur of professional successes and life misfortunes, both good and bad. On his return to Germany, he set up his own independent research group at the University of Berlin and made positive progress in his field of interest, phytochrome research. Particularly during the two years before the outbreak of the First World War, his research was very fruitful. Not only did he understand what chlorophyll was and how it worked in plant cells, but he also studied animal and plant health. His work on hemoglobin in animal cells won him the Nobel Prize in Chemistry in 1915. Despite this success, he ended his research career early, at a time when anti-Semitism was rife, and retired at the age of 53. Apart from a few students, he rarely interacted with others. The structure of chlorophyll was solved in 1940 by Hans Fischer, another Nobel Prize-winner in chemistry. However, the prize he won was ten years ago, for solving the structure and function of another important chemical, hemoglobin.

chloroplast

chlorophyll

Richard Willstätter

Hans Fischer

Melvin Calvin

At the same time, Robert Hill from the United Kingdom discovered that photolysis and CO_2 fixation in plant cells were two separate events. This was followed by Melvin Calvin's more detailed study of the second event, for which he was awarded the Nobel Prize in Chemistry in 1961. Calvin was born on April 8, 1911, in a Russian immigrant family, St. Paul, Minnesota. He graduated from the Michigan School of Mines and Technology with a degree of bachelor of science in studying chemistry at the age of 20, and four years later earned a PhD in chemistry from the University of Minnesota, before embarking on his own independent research journey at the University of California, Berkeley, at the age of 36. He had already demonstrated his love of research during his doctoral studies, which focused on electron affinity studies of a class of compounds called halogens. After graduating, he spent two years as a postdoctoral researcher at the University of Manchester in the UK. His research into the synergistic catalysis of a wide range of substances further stimulated his development. After setting up his own group, he initially focused on the structure and activity of organic molecules. He gradually integrated these into his doctoral and postdoctoral work, influenced by the research trends of the time. Finally, in 1954, he succeeded in opening the black box of photosynthesis in plant cells by using the Carbon-14 isotope to label CO_2. Since then, his research interests have gradually shifted from chemistry to biology. As a result, half of the researchers in his laboratory work in chemistry and the other half work in biology.

In the same year, Daniel Arnon from the United States discovered how plant cells convert light into the energy substance ATP. This process is similar to the oxidative phosphorylation of mitochondria in animal cells and is called photosynthetic phosphorylation. It was Johann Deisenhofer, Robert Huber, and Hartmut Michel who solved the riddle of photosynthesis. The three of them worked together to resolve the three-dimensional spatial structure of the protein complex at the heart of the photosynthetic reaction in bacteria. For this, they were jointly awarded the Nobel Prize in Chemistry in 1988. From this point of view, the ability to harness light energy does not seem to have been patented by plant cells, which are rich in chloroplasts and can use light extremely efficiently for the conversion of energy and matter. Other organisms, such as certain bacteria and algae, have similar functions. As with the controversy about the origin of mitochondria in animal cells, the origin of chloroplasts in plant cells has been hotly debated. Most evidence to date points to the bacterial symbiosis theory. The presence of genetic material in chloroplasts and its homology to that in bacteria supports this possibility, and the discovery of photosynthesizing bacteria has further confirmed the link.

Daniel Arnon

Johann Deisenhofer

Robert Huber

Hartmut Michel

photosynthetic phosphorylation

As we have just mentioned, one of the important features that distinguishes animal cells from plant cells is the absence of chloroplasts. However, there is always an exception to this rule. There is an animal called the sea slug whose cells also contain chloroplasts. As a result, like plants, it can eat and drink simply by basking in the sun. The sea slug is a mollusk that lives in the sea and is also known as the sea hare due to a pair of protruding and slightly stubby tentacles on its head. Although it is a mollusk, it is a member of the crustacean family, along with snails and clams. Its shell ends up being a thin layer for evolutionary reasons. Sea slugs live mainly on the bottom of the sea because they like to use their large head to dig in the sand. It is easier to find them in sandy areas, and when they are hungry, they will find some small invertebrates to eat. They also live and feed on seaweed in areas where it is abundant. To avoid their natural enemies, they change color to match the color of the seaweed they eat. There is a species of green leaf sea slug that can not only change color after eating seaweed, but also incorporate the chloroplasts in seaweed into their own cells and use them for their own purposes. When other food is scarce, they can simply rely on these chloroplasts and then find a place to bask in the sun without starving to death. The amazing thing, of course, is not that they swallow the chloroplasts from plant cells directly into its own cells, but that they steal the special genes from green algae and fuses them with its own genes. Eventually, they resynthesize the components needed for chloroplasts and chlorophyll. Not many animals on Earth are able to do this, and sea slugs were the first to be discovered with this special ability.

Normally, when an animal eats a plant, the genes in the plant cells are not transferred to the genetic system of the animal cell. However, scientific discoveries often break this so-called common sense. In addition to the case of the sea slug mentioned above, another notable example is the discovery by Professor Zhang Chenyu of Nanjing University, who led a team in 2011 to discover that a tiny sequence of nucleotides unique to plant cells can enter animal cells directly after being digested in the stomach, conducted the first study of the cross-border flow and function of such nucleotides. The discovery was not recognized by other experts in the field at the time, who considered it impossible based

on their own common sense. He was rejected when he submitted his work to foreign journals, and it was only when the domestic journal *Cell Research* in China realized the importance of his discovery that he was able to report it. It was not until ten years later that foreign research groups verified his findings and realized their significance.

Although plant cells cannot move like animal cells and cannot repair damaged tissues as well as animals, the function of plant cells is no less than that of animal cells, especially from a developmental point of view. Each cell is an all-rounder. The most common example is the sprouting of new shoots from old branches. The so-called old branch refers mainly to those thick branches. The appearance of the bark is usually more mottled and cracked, and the leaves on the branch have withered. If you cut off such an old branch, you will find small shoots sprouting from the wound, just as seeds sprout from a small head in the soil in spring. Day by day, the shoots become newborn branches, covered with joyful green leaves. If every branch and leaf of a great tree were to be cut off, after a time, new branches would grow from the trimmed edge. They would emerge close together like willow leaves preparing green collars. From a distance, it would look like a pot placed in the position of an artistic bonsai. There would be upward, downward, and slanting branches. And far away, it would resemble a bird in a bare tree with one green nest after another.

It is clear that in plants, it is not only the seeds that can take root and germinate, nor only the tips of the branches that can grow. The whole body of plant cells, in case of need, can return to the earliest stage of development at different degrees. This is similar to the end of the development of cells in animals and going back to the embryonic stem cell stage. This regenerative capacity allows for regeneration that is not like the repair of animal tissues. Instead, it involves starting from the beginning of growth and development or generating a branch with leaves or growing into a complete plant. The most common example is the banyan tree, which grows in the southern China. The leafy banyan tree often grows into a forest, and its ever-extending trunk will grow roots in the air, forming aerial roots. Once it touches the ground in the process of drooping, the aerial roots will form the trunk again. This new trunk will then expand in all directions with the mother tree as the center. Here I quote from Huang Helang's *Banyan Trees in My Hometown*: "The two old banyan trees at the end of the bridge, one standing upright with lush foliage, the other growing into a strange S-shape, with a gnarled and sinewy trunk reaching diagonally into the stream, which we call the 'hunchbacked.' What is even more extraordinary is that this curved part of its heart was burned out, forming a groove more than a metre square, and yet it lived on, crossing the stream, raising its head and stretching its thick branches towards the blue sky." Another example is the green plant golden pothos that is often used in homes to accentuate the interior. No matter how small you cut it, even without its original roots, as long as there is a piece of stem and a leaf, and then plunge it into water, new roots will sprout from the stem in the water after a while.

So, what other uses are there for the intense plasticity of plant cells? We already know from the previous section how difficult it is for animal cells to revert from grandchild to grandparent. There is no shortage of scientists throughout history who have devoted their lives to this subject. Plant cells, on the other hand, can easily go back to their ancestors. Therefore, will cloning plant cells also be much easier? Exactly. Because of the plasticity of plant cells, cloning plant cells is very different from cloning animal cells. There is no need for the nucleus of one cell to be painstakingly injected into the cytoplasm of another cell using microtransplantation technology, nor do we need to borrow external forces such as viruses to hardwire potent genes into the cells. This can be achieved by placing the plant cells in the right culture medium. Because they come from the same plant, all the cells have exactly the same genetic characteristics. Of course, although all cells in the whole plant have high plasticity, there are differences in capacity in different parts of the plant. Usually, the cells in the roots are more suitable for cloning. The medium mentioned here is the same as that used in animal cell culture, except that it simulates the solid soil in which the plant grows.

Although the environment in which plant cells are grown does not require the same fine air filtration as animal cells, botanists still have to be careful about contamination. Usually, all that is needed is a relatively clean room with an ultra-clean bench, such as those used to grow animal cells. You are then ready to go. In addition, plant cells are relatively easy to obtain and handle. There is no need for enzymatic digestion of plant tissue; just a pair of scissors or a razor blade will suffice. The plant's rhizomes or tissues from wounds that tend to grow new branches (known professionally as healing tissues) can be cut into small pieces. Each piece is then placed directly into a bottle of solid culture medium to be considered complete. To prevent contamination, the researchers place a clear or translucent plastic film over the mouth of the bottle and tie it with a rubber band or string. The next step is to place these babies containing plant cell clones under a lamp and wait for the plant cells to take root inside the bottle.

Using plant cell cloning technology, researchers have carried out large-scale rescue, protection, and propagation of rare, endangered, medicinal, and economically important plants. In addition, the soilless solid culture technology can not only realize the cloning of plant cells and the tissue cells obtained from the isolation of a plant into a considerable number of seedlings, but also through further optimization of the culture medium, provide sufficient nutrients to form a liquid culture medium that allows these seedlings to continue thriving. Today, this technology has fully moved out of scientists' laboratories and into marketable applications, mainly for the production of vegetable plants that people need every day. It is conceivable that the future of plant farming will be based entirely on modern science and technology, a three-dimensional factory farming system that is no longer dependent on the field and natural climate. Agriculture is no longer a traditional industry that relies on the soil and weather for sustenance, but has become a modern industry based on high-tech and the integration of various technologies. Farmers are no longer uneducated cultivators, but composite talents who master botany, mechanics, optics, and information technology.

Research on plant cells is aimed, on one hand, at better conserving the diversity of plant resources and, on the other hand, at improving crops and fruits for human consumption. Especially the latter, which is the basis for human survival. For hundreds of years, humans have been conducting research to improve crops. In the early days, the focus was on developing hybridization techniques. As technology and our understanding of the cell advanced, polyploid breeding was introduced. What is polyploidy? We already know that both animal and plant cells have genetic material in their nuclei, and for different species and cell types, this material—chromatin—formed by the intertwining of chromosomes, all have a fixed number. The genes that control different plant traits are often located on different chromosomes. If chromatin containing genes for different traits is artificially integrated into different cells, or if chromatin containing genes for beneficial traits is amplified to create polyploid plant cells, the final plant will have even better traits than before.

Taking this a step further, skipping complex structures like chromatin and intervening directly in genes for key traits would be more efficient and convenient for crop improvement. This is why we see scientists sending seeds of different plants into space for space breeding. The main purpose is to test the effect of the space environment on plant development. Additionally, we will use the ubiquitous radiation in space to mutate the genes in the seed cells and make further trait observations on the randomly generated mutations to screen for plants with beneficial traits. In addition to irradiation, modern molecular biology can also be used to edit genes in plant cells. For example, deleting genes that control height in rice can make it shorter and less prone to falling over, thereby increasing yield. Adding pest resistance genes and health-promoting genes to rice can prevent rice pests. Producing rice with higher nutritional value is also possible. In terms of the development trajectory, the acquisition of transgenic plants is moving from natural selection to artificial selection. This is both a result of historical evolution and a trend for future development. Of course, we still need to take a long-term and comprehensive view of the impact of genetically modified crops on the natural environment and the safety of humans and other animals. However, we should not give up eating for fear of suffocation and slow down the pace of technological development.

All of the above transgenic manipulations are done from a breeding point of view, in order to obtain a more human-compatible class of plant genetic traits that are needed. Another type of transgenic manipulation, like that of animal cell engineering, is to use plant cells or even the plant itself as a bioreactor to produce substances in the plant that originally belonged to other organisms or animals, in order to obtain more of a particular type of protein or drug that can be used for therapeutic purposes. Compared to animal cell engineering, the use of plants as bioreactors is cost-effective, especially in terms of greatly reduced economic costs both in the production phase and in the later storage phase, plant cell engineering has shown great advantages. Since the 1990s, when scientists first succeeded in using tobacco to produce surface antigens of Streptococcus mutans, the use of genetically modified plants for the production of vaccines and drugs has grown tremendously. In addition to tobacco, a wide range of fruits and vegetables have joined the genetically modified army, including tomatoes, apples, and potatoes. The vaccines produced work in two main ways: either by extracting the antigens in the plants and then injecting them into the human body, or by directly consuming the plants containing the antigens and then producing antibodies. To date, transgenic plant vaccines have been produced in a number of areas, including bacteria, viruses, diabetes, parasitic diseases, and contraception. In addition, good experimental results have been obtained by expressing antibodies in plants, as well as a variety of active peptides, including interleukins, serum proteins, and insulin.

A Vision of the Cell's Future

$\mathcal{B}$ased on the current theoretical foundation and development trend of cell science, we have every reason to believe that artificial cells and organoids may be practically applied in the twenty-first century. Based on engineering production technology, after solving the problems of safety and output efficiency, the first cell to go into clinical application may be blood cells. With further refinement, the first would be platelets, followed by erythrocytes, and then other immune cells or stem cells, such as T-cells and hematopoietic stem cells. As the seed of blood transfusion components, artificial blood will definitely replace today's blood donation for the benefit of any patient in need of blood transfusion. Following the development of artificial blood cells, dopaminergic neural pro-genitor cells generated by differentiating pluripotent stem cells may be applied to the treatment of Parkinson's disease. A breakthrough may be achieved in the near future, allowing patients to no longer suffer from unsteady grasping of spoons and difficulty in self-feeding.

In addition, based on a deep understanding of organ development and tissue structure, combined with rapidly developing materials science, we can induce cells at an early stage of development to form organoids with complete structure and better function in a three-dimensional culture system under in vitro conditions. Using these organoids, we can replace traditional animal experiments, reduce the need for model animals, and carry out various experiments and tests. On the other hand, we can try to use organoids with lower complexity and development closer to normal tissues to carry out alternative transplants for damaged tissues or organs. The most mature and earliest organoid to be used may be skin tissue, followed by an artificial liver made by combining solid liver cells obtained in large quantities. An artificial cornea can be made by combining corneal cells differentiated from pluripotent stem cells, and a cardiac patch can be made by superimposing cardiomyocytes differentiated from pluripotent stem cells. Of course, there are still many scientific problems and technical difficulties to be overcome for each type of cell and organ. For example, how to restore the polarity of hepatocytes in an artificial liver and how to keep induced cardiomyocytes and patches beating at the same rate as the patient's heartbeat.

For complex tissues and organs that are difficult to form through natural cell development, scientists have had some success using 3D printing technology, such as 3D printed hearts. This technology uses cells as the printing ink, stacked on top of each other according to a pre-designed model. It also uses the main components of the extracellular matrix, such as collagen, as the glue to hold the cells together. However, this technology is only a preliminary demonstration of the prospect of the intersection of different disciplines, and its future application is still very young compared to the development of the previous two technologies. Key issues that need to be addressed include how to solve the problem of superimposing different cell types in an organ when current 3D printing technology essentially uses a single cell type, and how to minimize the damage to cells caused by mechanical manipulation and improve cell viability.

Although technological advances have brought us a step closer to tissue or organ regeneration by allowing us to come into close contact with different types of cellular therapeutic techniques, it remains one of the ultimate goals of cell biology and a dream of regenerative medicine to realize tissue or organ regeneration under in vivo conditions, just as we have been able to do with the miraculous organisms of nature mentioned above. At present, we are trying to study cell fate transition under in vivo conditions, stem cell activation, and the stimulation of quiescent cells to proliferate. This may give wings to this ultimate goal and dream. For example, the direct conversion of glial cells, which proliferate in large numbers during brain injury, into neurons, which are lost in large numbers during brain injury and cannot be recovered, may provide feasible therapeutic solutions for various types of neurological disorders. Although cardiomyocytes are virtually incapable of proliferating, if cardiomyocytes adjacent to infarcted tissue can be activated with drugs to enter a rapid proliferative state and replace dead cardiomyocytes, we may be able to provide a regenerative therapy for heart disease, which has a high incidence of morbidity.

Existing virus-mediated gene-editing techniques have shown great promise in modulating cell function and fate in vivo, but safety concerns remain. Therefore, the development of other intervention techniques is of particular importance. The development of optogenetics, which uses different wavelengths of light to modulate the activity of photosensitive proteins, is currently widely used in the field of neurology. As the technology matures, it is not difficult to imagine that it will flourish in other fields. However, current optogenetic techniques have significant limitations, particularly the need to insert optical fibers at the site of intervention, which is both traumatic and inconvenient. If magnetic fields, sound waves, or heat, which can penetrate tissues and have non-invasive properties, can be used to replace light for intervention, it will have a wider range of applications. Based on these ideas, related technologies have also been derived, such as the study of magnetogenetics and the study of heat-sensitive proteins. Although these studies are still very naive, they have at least shown the way forward.

For in vivo regeneration, there are also some scientists who are working on a completely different line of research: the xenotransplantation technique. As the name suggests, this involves transplanting organs from one species to another. As we know, the main reason why humans have always been humans, dogs have always been dogs, and rats have always been rats is due to genetic differences and the resulting huge reproductive mismatch and immune rejection. From this, it occurred to some scientists that if the barriers could be broken down and the rejection eliminated, then it would be possible to realize cross-species use of animal organs on humans. This would solve the problem of insufficient donors for organ transplants that has long plagued humankind. To this end, scientists have begun to add human cells to monkeys and pigs at an early stage of embryonic development, creating monkeys and pigs with human cell chimeras, which are, to some extent, heterologous hybrids. If it is possible to replace all the cells of an organ or tissue in an animal with human cells, it is certainly not a dream to use animals as bioreactors to produce organs for human transplantation. In addition, there are scientists who are trying to directly remove or modify the genes in animals that are susceptible to human immune rejection. This is done in order to obtain animal organs that can be transplanted directly into humans, thus realizing the dream of xenotransplantation.

Although in vitro fertilization, as described in the previous article, can artificially intervene in the process of human reproduction to some extent, freeing it from dependence on nature and the helplessness of reproductive disorders, it still relies on naturally developed sperm cells and eggs, as well as the uterus, which serves as an incubator for the development of the fertilized egg. With the development of new technologies, scientists can now take skin cells and "reprogram" them into powerful stem cells. These stem cells can then be transformed into many different types of cells, even sperm and egg cells! In addition, with advances in tissue engineering, scientists have developed artificial uteruses that can briefly replace the natural uterus and continue to support a full-term sheep fetus until birth. Artificial uteruses can even support the development of an early mouse embryo of only about 200 cells, which appears to be a chaotic mass of flesh, to stay in vitro for six or seven days until three different embryonic

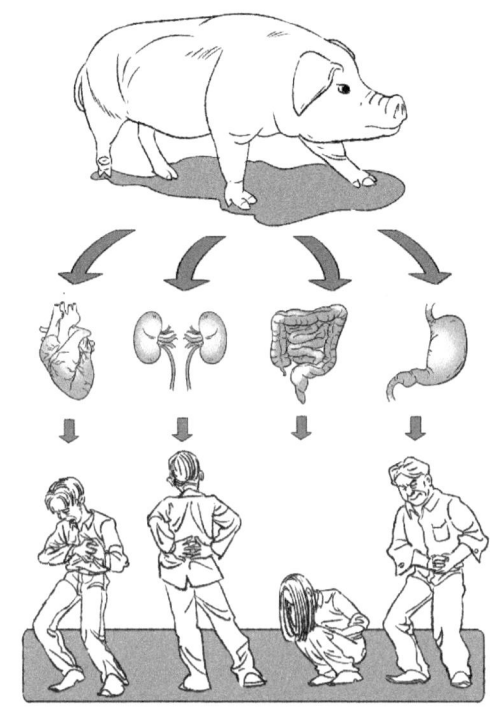

layers appear, resembling an early mouse form with a head, torso, limbs, and heartbeat. Based on the further maturation of the above two technologies and the research and development of human cells and the uterus, it is reasonable to believe that in the future society, it is possible not to rely on the natural sperm, egg, and human uterus for the continuation of offspring. People with reproductive deficiencies can choose to obtain artificial sperm and egg through skin, blood, or other kinds of cells. Then, after the artificial insemination is completed, they can continue to use the factory's artificial uterus for conception until the baby is born. This sounds very incredible and faces the huge challenges, especially in terms of legal, ethical, and other aspects. However, from the current state of theory and technological development trends, all this is highly feasible.

Because cells play such an important role, we need to establish cell banks for some of the cells of great therapeutic importance, such as blood banks, bone marrow banks, umbilical cord blood banks, sperm banks, egg banks, and

pluripotent stem cell banks. However, the various cell banks that have been established to date are essentially derived from donations from different human individuals. This is not a long-term solution, both in terms of economic cost and future development trends. Together with our mastery of cell production technology and our understanding of cell fate transitions, the complete replacement of cell donation by cell production technology will be the scenario of the future. In the future, we can imagine that a hematopoietic stem cell factory or bank will replace the current blood bank, bone marrow bank, and cord blood bank as an essential link in the future social and medical industrial chain. This is based on the ability of hematopoietic stem cells to self-replicate and produce a constant flow of blood cells.

With the deepening knowledge of cells and the wide application of cell therapy technology, people can develop other cell-based derivatives. The whole range of cells in the human body has great application value and significance. Therefore, the establishment of a cell bank for the whole range of cell types also seems inevitable. At present, there are private companies that have started to establish cell banks for immune cells. However, their aim is to operate and profit from them. From the point of view of human civilization and reproduction, the protection of human cellular resources and those of other species, as well as the establishment of a species-wide cell bank, are matters that require urgent international cooperation.

Both tissue repair and organ transplantation rely on replacing cells themselves, so is it possible to use non-cellular substances to mimic and replace cellular functions and thereby achieve tissue and organ regeneration? Because the cell is a highly integrated and delicate structure with multiple complex functions, replacement can only start from the simplest cell, step by step. Among them, red blood cells do not have a nucleus and their main function is to transport oxygen. Scientists have synthesized a variety of substances to help patients in desperate need of red blood cells. These substances may not look like red blood cells in form, but they can functionally bind, transport, and release oxygen. This is a success story. The synthesis of other cells has not yet been reported. However, based on the rapid development of synthetic biology technology, people have begun to try to synthesize some of the structures and organelles in cells, such as artificial cell membranes and synthetic chromosomes. As the technology matures and the demand for artificial cells increases, the direct synthesis of cells with various functions using chemical products and their direct application in regenerative medicine will be an epic invention for all of humanity. For plant cells, people are trying to synthesize chloroplasts. If they are successful and can be fused with animal cells, will animals be able to survive using only sunlight and air to breathe? For humans, they will become "vegetarians" in some sense.

Furthermore, if the corresponding nanorobots are designed to mimic different types of cells in form and function, then these real cellular robots can be used to replace cells in the organism and partially participate in the daily functions of certain tissues or organs. Of course, this will be a long road of scientific exploration, both theoretical and technical, with many difficulties to overcome. However, it is also an attractive avenue for future technology, especially in the case of tissue or organ deficiencies caused by acute injuries. In such cases, the use of this technology can achieve the goal of rapid repair and save those patients who cannot be treated by existing medical technology. Furthermore, if the function of these cellular machines is far better than that of normal organic cells, they have the potential to improve the function of the organism locally or systemically, realizing the so-called superpower. There are already technologies that can use physical electronic materials to simulate part of the tissue, such as nerve tissue, to string together broken nerve fibers, causing partially paralyzed limbs to regain nerve signals and the ability to move. Of course, the complexity of the microscopic world is far greater than that of the macroscopic organization. It will be no small challenge to move from macroscopic simulation to microscopic simulation, not only to achieve the imitation of the structure and function of each cell but also to allow each cellular machine to establish harmonious communication with each other. However, once it is achievable, it will be a revolutionary advance.

Both animals and plants, as well as cells themselves, depend on organic matter and have a certain life-cycle limit. However, when life transcends such material carriers, it seems that it can transcend the life-cycle limit and achieve immortality. At this point, some scientists have begun to try to download human consciousness and transfer it to computers and networks, so that even if the person dies, their mind will continue to exist on the network. Of course, this idea has been dismissed by many as a fantasy, especially since there is still a huge black hole in the current human understanding of consciousness itself. People haven't even figured out what the nature of consciousness is, so how can we talk about simulation? However, the cell is different. People's understanding of the structure and function of the cell has a hundred-year history. Research has moved from the macro level to the molecular level, and with the passage of time, a comprehensive analysis of the cell is just around the corner. For more than a decade, researchers have been trying to use computers to simulate the full function of a cell in order to create a virtual cell. To put it simply, although these cells can't carry out therapeutic applications in the practical sense, they can carry out all sorts of experiments on the cells, such as whether a certain drug is toxic to a certain type of cell, whether it promotes or inhibits cell growth, and so on. And all this can be done with just a computer or a mobile phone connected to the network.

Organic life collides with silicon-based computers. The virtual cell mentioned above is a combination of life on computers. What would it be like if the opposite were true, and computers were based on life? What would that look like? We know that computers are based on the repetition and combination of numbers 0 and 1 in their algorithms, whereas the genetic material in a cell

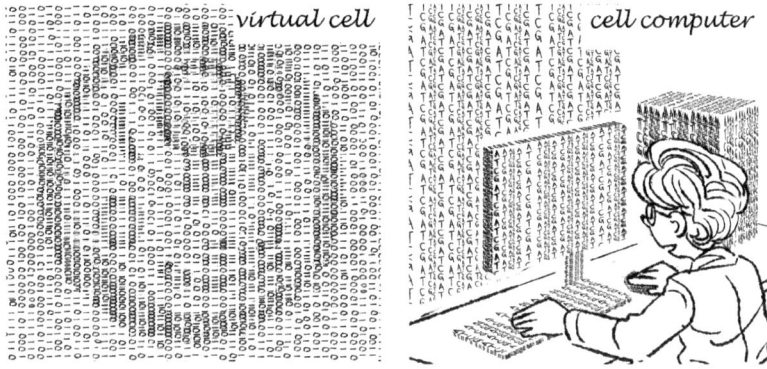

consists of the repetition and combination of four substances represented by the letters A, T, C, and G. The second combination is mathematically much more complex than the first, and it is possible to simulate the former with the latter. Mathematically, the second combination is much more complex than the first, and it is possible to simulate the first with the second. In this way, the code for the combination of 0 and 1 can be converted into an ATCG code. The required sequence can then be synthesized using conventional DNA sequence synthesis techniques or, if such information needs to be read, interpreted using sequencing techniques. By using the genetic material in cells to store computer information, it is possible not only to use a very small number of cells to store a large amount of data, but also to protect this data by replicating and freezing the cells for better distribution. If the use of cells to store data at will can be realized, then the next step should be to use a variety of fast and orderly chemical and biological reactions within the cell to carry out computational processing. It is particularly worth mentioning that the heat generated by the computer's rapid computation does not need to be bothered with by the cells, which can be regarded as one of the advantages of cellular computing. Of course, the cellular computer could replace today's silicon-based computers, and the future of quantum computers could be similar. Let's wait and see.

The Indispensable Microscope

$\mathcal{T}$o do a good job, an artisan needs the best tools. The progress of science cannot be separated from the development of technology. Like the discovery of the cell at the beginning of this book, without the invention of the microscope, the cell, as the most important unit of animal and plant organisms, might forever remain silent outside the vision of humankind. However, there is no accurate historical record of the first inventor of the microscope. Leeuwenhoek and Hooke can only be credited with improving and enhancing the early microscope in terms of magnification and ease of use. As a sibling of the microscope, the telescope was invented much more earlier. It was once thought that Galileo, a man of great ability who had used telescope to make essential discoveries, was the inventor of the microscope, but whether this is true remains to be proven.

Another Dutch optician, Zacharias Janssen, is considered the most likely inventor of the compound microscope. However, there are those who question this and believe that his father and another diplomatic friend, William

Zacharias Janssen

Boreel, also played an important role. This doubt comes from the record that the microscope bearing Jenssen's name in the Middelburg Museum in the Netherlands was made in 1595 when Zacharias Janssen was just 15 years old. This microscope was a simple device consisting of three circular sleeves. Two of them could be moved in and out to focus the magnified object for a clearer view, up to a maximum magnification of ten times. Unlike modern microscopes with stands, microscopes of the time were handheld, similar to the monoculars we see in films or on television. Although the imaging and magnification of early microscopes were crude and limited, the invention of this instrument was a fundamental breakthrough and a historic starting point.

As for the translation of the word "cell," we already know Li Shanlan's contribution. So when was the word microscope translated into Chinese and who did it? As one of the words whose characters have both the same form and the same meaning in Chinese and Japanese, many people may think that the word was translated from Japan and then imported into China, like many other scientific terms. After all, at that time, in the general environment of learning from the West, it was very easy to let people mistakenly think so. According to ancient and modern Chinese and foreign historical records that can be accessed, the microscope in English originated from the Latin word "microscopium." The word "micro" means small, and "scopium" means to see and check. It was first seen in the description of Galileo's friends. There is no clear record of when the microscope was first introduced to China and by whom it was first translated. It can only be glimpsed in the many manuscripts handed down by the literati. The word "显微镜" (microscope, *xianwei jing*) had already appeared in the *History of Mirrors* and the *Cantonese Xinyu*, which can be traced back to the second half of the seventeenth century when the microscope was introduced to China. But why the word "显微" (*xianwei*) plus the word "镜" (*jing*) instead of "放大镜" (magnifiers, *fangda jing*)?* It is possible that the word "显微镜" was deliberately used to distinguish it from the word "放大镜," as "显微" was a phrase derived from the *Book of Changes* and means to reveal the subtle and clarify the hidden. It is interesting to note that the earliest recorded reference to microscopes in Japanese dates back to the early eighteenth century when the word was translated as "bug-eye glasses." Thus, the translation of the word "microscope" as "显微镜" was first proposed in China and then introduced to Japan. This is very much in line with the translation and spread of the word "细胞."

Back to the point, the most important technology for the creation and development of the microscope is the use of light and its recognized imaging principles. Although early microscopes were able to magnify microscopic objects, whether it was a few times in the beginning or tens of times in the later stage, they relied on the result of a single convex lens. Hence, they are also called single-lens microscopes. Such microscopes had the advantage of being simple and easy to make, but they also had the problem of limited magnification and blurred images. To solve these problems, compound microscopes were later developed. In these microscopes, several lenses were stacked on top of each other and a convex lens was used to further magnify the image formed by the previous lens. At the same time, however, a new problem arose, namely the existence

---

* The word "microscope" (显微镜) in Chinese is a combination of "reveal the micro" (显微) and "glass" (镜), which can be directly translated back to English as "the glass that is used to reveal the micro." Meanwhile, the word "放大镜" in Chinese consists of "magnify" (放大) and "glass" (镜), which can be explained as "the glass that is used to magnify." The author's question is, hence, based on the similarities and differences in their Chinese meanings.—Trans.

of spherical phase differences between different lenses, which seriously affected the imaging effect. This problem plagued the compound microscope for more than a century until it was greatly improved by an amateur microscopist, Joseph Jackson Lister. Born on January 11, 1786 in London, England, into a family of liquor merchants, Lister was forced to leave school at the age of 14 to learn the trade with his father. It was probably through his father's example that he learned how to be a successful businessman. In addition to the liquor business, he became involved in ship investments. With strong financial backing, he was able to develop his interests outside of his main business. This included participating in various activities organized by the church, where he was introduced to microscopy. Through independent research, he solved the microscope problem of the century by precisely adjusting the distance between each lens in a compound microscope to improve the clarity of the microscope image.

Joseph Jackson Lister

In the field of microscopes, one name that rings a bell is Carl Zeiss whose lenses represent the very best configurations. It is the name of a brand, company name, and a person. Carl Zeiss was born on September 11, 1816, in the small German town named Weimar, which is the birthplace of the world-famous poet Johann Wolfgang von Goethe, the famous opera singer Caroline Jagemann, and the musician Franz Liszt. In addition, Weimar was already an extremely prosperous trading town, creating a demand for luxury goods and a large number of craftsmen working in a variety of trades. In this environment, it is easy to see how Zeiss was able to build his own business empire. Microscopes were an absolute luxury at the time. It was not Weimar, however, but Jena, 20 kilometers from Weimar, that really put Zeiss on the microscope map. It was there that he met two other heavyweights who would rewrite the history of microscopy: Ernst Abbe and Otto Schott.

Abbe is a legendary figure who illustrates the importance of knowledge in changing one's destiny. He was born in Germany on January 23, 1840, into a poor family. His parents had to work 16 hours a day as weavers to support their family. Fortunately, even in these difficult circumstances, Abbe was lucky enough to go to school at the appropriate age like all the other kids. He was so gifted that he outperformed his peers from the start of primary school. When he was in fourth grade, his teachers advised him to move to a better school to avoid being held back. However, financial constraints made this difficult for his father. It was only when he received financial support from the owner of his father's factory that he was able to make the journey. Since then, Abbe had excelled in his studies, graduating from university with the help of various scholarships and earning a PhD. He had shown great interest in the fields of mathematics and physics. After graduating, just as expected, he stayed on as a university lecturer, married, and had children in his prime. His stable life allowed him to develop hobbies outside of work. It was during this time that he developed a friendship with Zeiss and used his skills to help him improve conventional microscopes. This contribution played a role in the birth and growth of Carl Zeiss AG.

Carl Zeiss   Ernst Abbe   Otto Schott

After Zeiss himself died, Abbe took over the company and not only set up a foundation to run the business but also carried out drastic reforms to the company's management, including a series of popular measures such as paid holidays, the eight-hour workday, employee stock ownership plan, and pension schemes. This laid the foundations for Carl Zeiss AG to grow into a business empire. Abbe made great achievements not only in business but also in science. The most important discovery was Abbe's limit theory, which is the formula for the resolution limit of microscopes.

With the platform built by Zeiss and the foundation of Abbe's mathematical theory, high-quality glass was the last step to be taken to revolutionize early microscopes. Just as good iron is needed to make good steel, even a skilled cooker can't cook without rice. It was Schott's turn. Schott was born on December 17, 1851, in Witten, Germany. Although he was the youngest compared to the first two, he had his own advantages. Schott's father was a glassmaker, and Schott developed a keen interest in the industry from an early age. This led him to study chemistry, metallurgy, and physics throughout his education, culminating in a doctorate. At the age of 28, he developed a glass containing lithium that had different optical properties than conventional glass. When he wrote to Abbe, who was well-known in the field, they began a close relationship and collaboration. Three years later, he joined Abbe and Zeiss to systematically test the optical properties of glasses containing different elements and their use in microscope. The formation of the golden trio revolutionized microscopy, making Carl Zeiss AG the leading microscope manufacturer and a mythical presence in the world of optical lenses. Schott himself became the founder of modern glass manufacturing.

Golden trio's first revolutionary product was the compound achromatic microscope. This led to a good and professional atmosphere, which in turn attracted more talented people to join the Zeiss company and develop even more outstanding products. First, Paul Rudolph designed the first achromatic microscope. He was followed by August Köhler, who built the first ultraviolet microscope. But even an old horse can occasionally lose its footing, and Carl Zeiss AG missed the mark twice, both times in favor of the Nobel Prize in Physics. The first misstep was the super-resolution microscope, based on Richard Zsigmondy's discovery of the phenomenon of heterogeneity in colloidal solutions. Zsigmondy's early work focused on the study of color on glass or ceramic surfaces. As a result, he crossed paths with Schott and was employed by another company where Schott once worked until the early 1900s. It was during this period that he began to study the chemistry of colloids, for which he was awarded the Nobel Prize. He applied colloidal preparation techniques to the preparation of microscope specimens, contributing to the introduction of the slit microscope. The second missed product was the phase contrast microscope, which is unique in that it uses the phase change of the sample to create mutual interference of light, allowing clear observation of cell outlines and internal structures. If cells were viewed with a conventional microscope, they would have to be stained before they could be imaged. The inventor of this Nobel Prize-winning achievement is the great name in the field of optics, Frits Zernike. Although famous, when he wanted to develop a new type of microscope based on his theory of phase contrast of light in the early days, Carl Zeiss AG scoffed at it until ten years later. With the financial support of others, his idea was realized and later gained wide recognition and application. Although Carl Zeiss AG achieved an absolute market share by virtue of its excellent craftsmanship, it shied away from innovation and missed out on one epoch-making microscope product after another.

The birth and development of the above-mentioned microscopes, although more advanced from one generation to the next, is still based on the overall

Richard Zsigmondy

Frits Zernike

scientific logic of the use of light. In the range of visible light, as the wavelength of light gradually decreases, the magnification gradually increases. However, according to the Abbe's limit theory, when the wavelength limit of visible light is reached, the magnification also reaches its limit. At this point, other physicists have studied acousto-optic electromagnetism to its very essence and have successively proposed various theories, including the wave-particle duality of light and the electromagnetic transition. If the magnification of the microscope was to be increased, the wavelength of the incident light had to be shortened even further. It was on the basis of these researches and theories that electrons, whose wavelength is much smaller than that of any visible light, came to mind. Max Knoll was the early beginning of related research pioneers, but the characteristics of the electron are very different from traditional light. Methods to produce electrons and control the direction of their propagation are required to solve the problem. Fortunately, Knoll recruited Ernst Ruska, a gifted student in this field, and together they finally built the first electron microscope in 1931. Just as the first computer in history was the size of a room and had mediocre computing power, the first electron microscope had a maximum magnification of only 17 times. Two years later, they were able to increase the magnification to more than 1,000 times, leaving the best optical microscopes of the time in the dust. First, the object magnified by the electron microscope was a metal grid. Since Ruska's brother was in biomedical research, they soon extended the use of the electron microscope to life science outside of physics, thus unlocking the black box inside the cell. As described in previous chapters, one electron microscope observation and discovery after another was competed for, laying the foundations for modern cell biology research. And in this round of competition, the German company Siemens came out on top, taking the lead in supporting Ruska's research, iterating generations of electron microscopes, and becoming the leader in the field.

electronic microscope

Max Knoll

Ernst Ruska

Gerd Binnig

Heinrich Rohrer

The next step, the introduction of the electron microscope, seems to break the usual way and go beyond people's imagination. Can you imagine a person with no eyes but can see things, no ears but can hear the sound, and no nose but can smell flavor? This microscope has gotten rid of the reliance of conventional microscopes on light and electrons, but the use of an extremely small probe to detect the surface of the object, like the blind use crutches to explore the road. When the distance between the probe and the object changes, the voltage between the two causes electrons to move, resulting in a change in current. By detecting this change in current, the shape of the object can be mapped. The resolution of this detection technique reaches the atomic level. The inventors of this marvelous microscope, known as the electron scanning microscope, are Gerd Binnig and Heinrich Rohrer, two employees of IBM in Switzerland. Although their invention was born in 1981, it was not until five years later that they were awarded that year's Nobel Prize in Physics, together with Ruska. However, Rohrer had already died and was unable to accept the honor.

Since its birth, the electronic microscope has been destined to change the course of human science and has attracted countless heroes to compete for it. Due to the complexity of the electronic microscope's construction, which is subject to many disturbing factors, even a small drop of water can cause a qualitative change in the imaging and magnification performance of the object. Jacques Dubochet from Switzerland has spent his life researching for such a drop of water, so that the power of the conventional electron microscope can leap up another level.

Born on June 8, 1942, in the wine-producing town of Aigle, Switzerland, during the German invasion of Moscow and the Nazi siege of Switzerland, Dubochet has spoken frankly that he is grateful for his optimistic parents. Otherwise, he would not have been conceived and born by his mother. He is also grateful for the many teachers he had during his early school years. At the age of 15, he was diagnosed with dyslexia and his academic performance went downhill for a while. He had patient teachers who did not give up on him. They spent a lot of time with him to help him overcome the disorder, even encouraging him to give a speech in front of the whole class to keep him from dropping out of school. Even so, his developmental disabilities affected his academic and social skills to some extent. At the age of 20, he was able to enter university thanks to the life skill training from his elder sister at home. As his father was an engineer, he naturally studied physics at university. At the time, physics was widely used in biological research, and influenced by one of his favorite university teachers, he chose to specialize in biophysics for his PhD. There, he was introduced to electron microscopy. Thirty-seven years after the invention of the electron microscope, the European Molecular Biology Laboratory had just opened ten years later in a wooded area in Heidelberg, Germany. The beautiful surroundings and the new atmosphere of scientific discovery attracted many scientists to the laboratory, and Dubochet was one of them. Based on his own research expertise, he began to study the treatment of water during sample preparation for electron microscopy, a seemingly obscure area of research. After nearly 30 years, he finally succeeded in solving the shortcomings of water in the preservation of cellular samples by vitrifying the water to maximize the preservation of the natural state of the various samples inside the cell. This solution led directly to his collaboration with Joachim Frank and Richard Henderson, and the development of the cryo-electron microscope. The three were jointly awarded the 2017 Nobel Prize in Chemistry for their work. Why is it a chemistry prize and not a physics prize? The invention of cryo-electron microscopy directly accelerated the resolution of the three-dimensional structure of proteins. This did not only increase the resolution by an order of magnitude, but also enable the resolution of many protein structures that could not be resolved by traditional solutions. The team led by Professor Shi Yigong from China made outstanding contributions in this regard.

Joachim Frank

Jacques Dubochet

Richard Henderson

Although the performance of the electron microscope far exceeds that of the optical microscope, each has its own advantages. Very often, the application of the optical microscope is still irreplaceable, and its application scenarios are also ubiquitous in modern biomedical research. Therefore, the iteration and research based on the optical microscope itself is still going on continuously. Although Abbe's limit theory was to some extent proposed to guide the direction of microscope development and led to the invention of the electron microscope, it also limited people's thinking. The theory reigned for more than a century before it was broken by Eric Betzig, Stefan Hell, and William Moerner. They brought the resolution of optical microscope up to the level of the electron microscope in the visible light range, giving rise to super-resolution fluorescence microscope. For this achievement, the trio won the 2014 Nobel Prize in Chemistry.

Although the three won the same prize, it was not the result of their joint efforts, but of their own seperately. The reason for this is largely due to the very different family backgrounds and upbringings of the three, and therefore their stories are very different. The three of them, one from the middle class and one from a poor family, with the last from a rich household.

Eric Betzig

William Moerner

Stefan Hell

Beziger was born on January 13, 1960, in Ann Arbor, southeast Michigan, the United States. His father engaged in wrestling coaching in the early years of the university and then, due to economic pressures, began a business. With tireless efforts from individuals, he went from being a small machine tool draftsman to a business owner with hundreds of employees. And all this, Beziger sees in the eyes, remembers in the heart, and also teaches him how to study and work hard. When he was in primary school, he was influenced by the father of a classmate who was involved in scientific research. As a result, he began to make contact with science and has loved it ever since. He often tinkered with all sorts of small scientific experiments. He graduated from high school and went on to study physics at university, where his hard work kept his grades at the top of his class. After graduating, Betzig joined Cornell University and the famous Bell Lab, where he became involved in microscopic imaging, which became his lifelong hobby. However, the favorable conditions also meant greater competition, forcing him to do excellent work or risk not only losing his job but also feeling very embarrassed. During this time, he often arrived at the lab at four o'clock in the morning to start work, and it was good to have a good friend who was as enthusiastic as he was about working with him. In the mornings, when one of them arrived first, the other would feel the temperature of the other's car to estimate how long he had been there. Hard work always pays off, and eventually he developed a new microscope with a higher magnification resolution than existing microscopes. Although it was a preliminary prototype and had little practical application, it showed great promise. It was all just beginning when Bell Labs was forced to close its doors for financial reasons, and Beitzig lost his job and became a stay-at-home dad. Staying at home, after a while he went to his father's company to help with odd jobs. During his idleness, he casually designed a prototype of a so-called modern hybrid car. However, no one asked for it at the time and it was never finished. After a long period of time, he became so frustrated that he decided to return to his old profession and read scientific literature every day. This actually allowed him to find a new way to further improve the microscope he had previously invented. The key to this method lies in fluorescence. After contacting a group of old friends in the academic world to verify this new idea, it was validated and a super-fluorescence microscope was born in the living room of his home. From this, we can see that scientific research does not depend on the size of the team. Even one person can do it. The key is to find the point of interest and be good at observing, thinking, analyzing, and summarizing. Of course, it would be even better if we could collaborate with others. As the saying goes, "A good man has three helpers."

In contrast to Betzig, Hell's life, as his surname "Hell" suggests, has been marked by many hardships. Born on December 23, 1962, in the city of Arad in western Romania, close to Hungary, Hell learned from an early age to be cau-

tious and maintain skeptical values due to the complex social circumstance at that time. He followed his parents around until he was 15 when the family emigrated to Germany. There, he was able to stabilize himself. The new environment allowed him to continue his education, but the superficiality of the books and rote learning made him bored and tired. However, later on, his university studies on microscopy gave him a taste for science and experimentation. Under the guidance of his mentor and the tasks assigned to him, the direction of his research was to address the shortcomings of the confocal microscope, which was in its infancy at the time, and improve its performance. However, the good times did not last long. The research funds were soon cut off, and he could only find another way to apply for new research fund support. In such intermittent financial support, the research project experienced a moment of stagnation and a moment to move forward. One round after another, the whole experience of nearly ten years of frustration led him to create the high-resolution fluorescence microscope prototype. This was not favored by the people of that time. If it were not for his own persistence in a long period of lack of financial support conditions, many people would have difficulty continuing. It was not until the beginning of the twenty-first century that he gradually gained recognition in the industry and established his own independent and stable research team. This team gradually brought the early prototype microscope to maturity and application.

Although Betzig and Hell independently developed visible light-based super-resolution microscope from different perspectives, they shared a common theoretical basis: single-molecule fluorescence. Without this early theoretical discovery, the two would have done nothing. Even with the best ideas, it would have been very difficult to break through the optical imaging limit proposed by Abbe. The person who put forward this theory was none other than Moerner. He was born on June 24, 1953, in Pleasanton, California, the US, in a military family environment. This allowed him to grow up with food and clothing, as well as the strict discipline of his parents. As a result, he developed a positive and enterprising nature. In high school, he won awards for participating in the school's science competitions. In his spare time, he even attended the university's summer science camps. Although Moerner studied electrical engineering at university, he switched to low-temperature solid state physics in graduate school, where he had the privilege of studying under several Nobel laureates. It was the relaxed environment and freedom to explore that allowed him to put forward the theory of single-molecule fluorescence shortly afterward, even though at the time he had no experimental support and could see no practical application. It is this kind of wild and seemingly useless knowledge that inadvertently shapes our science and technology today, and profoundly affects our daily lives.

Finally, the manufacture of microscopes must be mentioned. Throughout its history, the invention and development of the microscope has been virtually devoid of Chinese contributions. Of course, in the latest super-resolution fluorescence microscopy, Chinese scholar Zhuang Xiaowei has made certain contributions. However, her experiments were also carried out abroad. Historical limitation is an unavoidable factor, but with the development of the domestic economy and the overall progress of scientific strength, if we still rely on microscopes provided by foreign countries to carry out relevant research, we will certainly be subject to the monopoly of foreign technology in the field of life medicine. Of course, whether it is Carl Zeiss AG, IBM Lab, Siemens, or Bell Lab, they are accompanied by the strength of the country in which they have been able to grow up. This growth did not happen overnight, and the use of commercial funds to embrace the development of talent as well as the promotion of iterative technology is an important factor in casting their brilliant brand. Back to today's China, through the history of science, to find out the experience and laws, combined with the actual situation in the country, to find out a series of innovation and entrepreneurial road suitable for our development, is bound to be the only way out of domestic microscopes and even other important scientific instruments.

Epilogue

$\mathcal{I}$t took nearly two years from the time of writing to the completion of the manuscript, and when I put the pen to paper, I felt relieved. I think there are three reasons for this: first, as a teacher, I did not go back on my word; second, as a son, I was able to dispel my father's doubts; and third, as a father, I was able to answer my children's questions.

The writing of this book is partly because I have to commute three hours to and from work every day, especially the two hours on the subway. This gives me enough time to observe things and be quiet for ideas. Of course, I will use my cell phone from time to time to record a few words and look up some information. After writing this book, the greatest feeling is that many major discoveries in this field were often accomplished under very humble conditions, especially during the First World War and the Second World War when many scientists were either underfed, underclothed, or displaced. In such difficult times, they still managed to make discoveries that were written into history, and they deserve our respect. On the other hand, our current research and living conditions are far better than theirs, but the number of important discoveries seems to be much smaller, which is really a shame. Another thing that struck me was that by learning about the history of these historical figures, especially their teenage lives, it was clear that they all had a love for nature. Although many of them were mediocre in their studies when they were young, this did not hinder their development. On the contrary, as they grew older and gained more experience, their interest and resilience increased. They were able to overcome difficulties, understand the nature of the problem, and finally achieve results that were recognized by others. It is worthwhile for us to learn from them.

After discussing a brief history of the cell spanning three hundred years, the next topic worthy of scientific imagination may be the extent to which the development of cell science will impact human beings in the next three hundred years.

This is my first time writing a popular science book. Due to the many knowledge points involved, it is inevitable that there are mistakes and omissions. I implore you to criticize and correct them.

References

Andrea, A. J. *World History Encyclopedia*. Santa Barbara: ABC-CLIO, 2011.

Blundell, J. "Observations on Transfusion of Blood." *Lancet* 12, no. 302 (1829): 321–324.

Brownlee, C. "Biography of Rudolf Jaenisch." *Proc Natl Acad Sci USA*. 101, no. 39 (2004): 13982–13984.

Carreyrou, J. *Bad Blood: Secrets and Lies in a Silicon Valley Startup*. New York: Knopf, 2018.

Chen, Hui. *Approaching Wang Zhenyi*. Shanghai: Shanghai Jiao Tong University Press, 2011.

Croft, W. J. *Under the Microscope: A Brief History of Microscopy*. Singapore: World Scientific Publishing Company, 2006.

Davis, R. L., H. Weintraub, and A. B. Lassar. "Expression of a Single Transfected cDNA Converts Fibroblasts to Myoblasts." *Cell* 51, no. 6 (1987): 987–1000.

Diamantis, A., E. Magiorkinis, and G. Androutsos. "Alfred Francois Donné (1801–1878): A Pioneer of Microscopy, Microbiology, and Hematology." *J Med Biogr* 17, no. 2 (2009): 81–87.

Dolgin, E. "Bioengineering: Doing without Donors." *Nature* 549, no. 7673 (2017): S12–S15.

Gluckman, E., H. A. Broxmeyer, and A. D. Auerbach, et al. "Hematopoietic Reconstitution in a Patient with Fanconi's Anemia by Means of Umbilical-Cord Blood from an HLA-Identical Sibling." *N Engl J Med* 321, no. 17 (1989): 1174–1178.

Guo, Xiaoqiang. "Georges Koller." *Genetics* 31, no. 9 (2009): 873–874.

Hao, Yuping, Lu Lin, and Yang Zhihong. "Progress of Transgenic Plant Vaccines." *Journal of Nuclear Agriculture* 34, no. 12 (2020): 2708–2724.

Henig, R. M. *Pandora's Baby: How the First Test Tube Babies Sparked the Reproductive Revolution.* Annotated Edition. New York: Cold Spring Harbor Laboratory Press, 2006.

Hirsch, T., T. Rothoeft, and N. Teig, et al. "Regeneration of the Entire Human Epidermis Using Transgenic Stem Cells." *Nature* 551, no. 7680 (2017): 327–332.

Kampen, K. R. "The Discovery and Early Understanding of Leukemia." *Leuk Res* 36, no. 1 (2012): 6–13.

Kerr, J. F. "History of the Events Leading to the Formulation of the Apoptosis Concept." *Toxicology* 181–182 (2002): 471–474.

Kerr, J. F., A. H. Wyllie, and A. R. Currie. "Apoptosis: A Basic Biological Phenomenon with Wide-Ranging Implications in Tissue Kinetics." *Br J Cancer* 26, no. 4 (1972): 239–257.

Kolata. *Cloning: The Road to Dolly and the Path Ahead.* Shanghai: Shanghai Science and Technology Press, 2000.

Li, Shuxue, and Tang Junying. "Tong Dizhou: The Founder of Chinese Experimental Embryology." *Dialectics of Nature Newsletter* 42, no. 6 (2020): 120–126.

Liu, Rui. "Discovery and Origin Hypothesis of Mitochondria." *Teaching Biology* 41, no. 11 (2016): 9–11.

Liu, Ying, and Jiang Xia. "Research Progress of Cell Therapy for Diabetes Mellitus." *Chinese Journal of Cells and Stem Cells (Electronic Edition)* 7, No. 1 (2017): 59–63.

Ma, Qianhong, and Huang Zhongying. "Discussion on the New Concept of Embryo Freezing and Thawing Transfer Strategy in Assisted Reproductive Technology." *Journal of Obstetrics and Gynecology* 36, no.4 (2020): 243–245.

Mandai, M., A. Watanabe, and Y. Kurimoto, et al. "Autologous Induced Stem-Cell-Derived Retinal Cells for Macular Degeneration." *N Engl J Med* 376, no. 11 (2017): 1038–1046.

Niklason, L. E., and J. H. Lawson. "Bioengineered Human Blood Vessels." *Science* 370, no. 6513 (2020): eaaw8682.

Pan, Deng, and Gao Hongbo. "Discovery of Chloroplast DNA." *Biological Bulletin* 47, no. 7 (2012): 53–55.

Parker, A. *The Stem Cell Hope: How Stem Cells Medicine Can Change Our Lives.* Shanghai: Shanghai Education Press, 2015.

Pearce, J. M. "Rudolf Ludwig Karl Virchow (1821-1902)." *J Neurol* 249, no. 4 (2002): 492–493.

Picoult, J. *My Sister's Keeper.* Washington: Washington Square Press, 2005.

Purrington, R. D. *The First Professional Scientist: Robert Hooke and the Royal Society of London.* Basel: Birkhäuser Verlag AG, 2009.

Reed, J. C., and B. J. Druker. "Peter C. Nowell (1928—2016)." *Proc Natl Acad Sci USA* 114, no. 18 (2017): 4569–4570.

Rosenbaum, L. "Tragedy, Perseverance, and Chance—the Story of CAR-T Therapy." *N Engl J Med* 377, no. 14 (2017): 1313–1315.

Rous, P., and F. S. Jones. "A Method for Obtaining Suspensions of Living Cells from the Fixed Tissues, and for the Plating Out of Individual Cells." *J Exp Med* 23, no. 4 (1916): 549–555.

Sanford, K. K., W. R. Earle, and G. D. Likely. "The Growth in Vitro of Single Isolated Tissue Cells." *J Natl Cancer Inst* 9, no. 3 (1948): 229–246.

Savitt, T. L., and M. F. Goldberg. "Herrick's 1910 Case Report of Sickle Cell Anemia." *JAMA* 261, no. 2 (1989): 266–271.

Sipp, D., P. G. Robey, and L. Turner. "Clear Up This Stem-Cell Mess." *Nature* 561, no. 7724 (2018): 455–457.

Skloot, R. *The Immortal Life of Henrietta Lacks.* New York: Crown Publishers, 2010.

Snyder, L. J. *Eye of the Beholder: Johannes Vermeer, Antoni van Leeuwenhoek, and the Reinvention of Seeing.* New York: W. W. Norton & Company, 2016.

Takahashi, K., and S. Yamanaka. "Induction of Pluripotent Stem Cells from Mouse
Embryonic and Adult Fibroblast Cultures by Defined Factors." *Cell* 126, no. 4
(2006): 663–676.

Nobel Prize. https://www.NobelPrize.org.

Thorborn, A. L. "Alfred François Donné, 1801–1878, Discoverer of Trichomonas
vaginalis and of Leukaemia." *Br J Vener Dis* 50, no. 5 (1974): 377–380.

Wang, Liqun. "CAR-T and Immune Cell Tumor Therapy." *Chinese Journal of Cell
Biology* 41, no. 4 (2019): 537–548.

Xu, Kewei. "Formation of the Word 'Microscope' and Its Linguistic and Cultural
Exchange between China and Japan (1646–1831)." *Higher Japanese Language
Education*, no. 1 (2018): 137–148.

Xu, L. Wang J., and Liu Y., et al. "CRISPR-Edited Stem Cells in a Patient with HIV
and Acute Lymphocytic Leukemia." *N Engl J Med* 381, no. 13 (2019): 1240–1247.

Yang, Jianmin, Hao Huili, and Wang Weibin, et al. "Progress in Single Mammalian Cell
Encapsulation." *Chinese Science (Life Science)* 50, no. 4 (2020): 406–426.

Zhai, Zhonghe, Wang Xizhong, and Ding Mingxiao. *Cell Biology.* 4th ed. Beijing:
Higher Education Press, 2011.

Zhang, H. "Origin of the Chinese Word for 'Cell': An Unusual but Wonderful Idea of a
Mathematician." *Protein Cell* 12, no. 9 (2021): 671–674.

Zheng, Ruizhen. "The Scientific Life of Tong Dizhou." *Chinese Journal of Cell Biology*
41, no. 4 (2019): 774–784.

Zhu, Wenbing, Lu Guangxiu, and Fan Liqing. "The Establishment of Sperm Banks and
the Ethical Issues They Face." *Journal of Peking University (Medical Edition)* 36, no.
6 (2004): 670–672.

Index

ABOUT THE AUTHOR

Cheng Lin is a professor and a doctoral supervisor at Ruijin Hospital, Shanghai Jiaotong University School of Medicine. His research primarily focuses on cell reprogramming and its implications in disease development as well as its application in tissue regeneration, involving creating bioinformatics methods to understand cellular reprogramming processes, both in research and therapeutic contexts, particularly related to cancer.